KV-341-438

ENERGY CONSERVATION AND ENERGY MANAGEMENT IN BUILDINGS

Proceedings of the Conference organised by the Construction Industry Conference Centre Limited in conjunction with the Institution of Heating and Ventilating Engineers, the Illuminating Engineering Society, the Institute of Fuel and the Department of Energy

ORGANISING COMMITTEE

CHAIRMAN: A. F. C. Sherratt

N. F. Bradshaw, C.Eng., F.I.H.V.E. (*Institution of Heating and Ventilating Engineers*)
L. C. Bull, C.Eng., F.I.Mech.E., F.I.H.V.E. (*Institution of Heating and Ventilating Engineers*)
D. J. Croome-Gale, M.Sc., C.Eng., M.I.H.V.E., M.Inst.P., M.A.S.H.R.A.E., A.M.Inst.F. (*Institution of Heating and Ventilating Engineers*)
J. C. Denbigh, C.Eng., M.I.C.E., M.I.Mech.E., M.Inst.F., M.I.H.V.E. (*Department of Energy*)
N. J. D. Lucas, M.A., Ph.D., C.Eng., M.Inst.F. (*Institute of Fuel*)
P. L. Martin, C.Eng., P.P.I.H.V.E., F.I.Mech.E., F.Inst.F., A.M.R.Ae.S., M.Cons.E. (*Institution of Heating and Ventilating Engineers*)
W. A. Price, C.Eng., B.Sc.(Eng.), M.I.E.E., F.Illum.E.S. (*Illuminating Engineering Society*)

Energy Conservation and Energy Management in Buildings

Edited by

A. F. C. SHERRATT

B.Sc., Ph.D., C.Eng., F.I.Mech.E., F.I.H.V.E., M.Inst.R.

Assistant Director, Thames Polytechnic, London, England

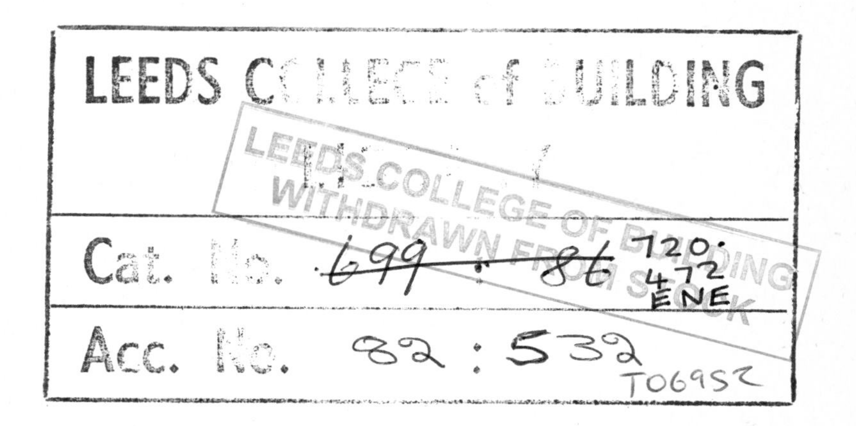

APPLIED SCIENCE PUBLISHERS LTD

LONDON

APPLIED SCIENCE PUBLISHERS LTD
RIPPLE ROAD, BARKING, ESSEX, ENGLAND

ISBN: 0 85334 684 4

WITH 50 ILLUSTRATIONS AND 38 TABLES
© CONSTRUCTION INDUSTRY CONFERENCE CENTRE LIMITED 1976
P.O. BOX 31, WELWYN, AL6 0XA, ENGLAND

All rights reserved. No part of this publication may be reproduced, stored in a retrieval system, or transmitted in any form or by any means, electronic, mechanical, photocopying, recording, or otherwise, without the prior written permission of the publishers, Applied Science Publishers Ltd, Ripple Road, Barking, Essex, England

Printed in Great Britain by Galliard (Printers) Ltd Great Yarmouth

Contributors

P. R. ACHENBACH, B.S.E.E., B.S.M.E., F.A.S.H.R.A.E.
Chief, Building Environment Division, Center for Building Technology, National Bureau of Standards, Washington, D.C., U.S.A.

K. R. ACKERMAN, C.Eng., B.Sc.(Eng.), F.I.E.E., F.Illum.E.S.
President, The Illuminating Engineering Society, Head of Television Planning, Studio Capital Projects Department, British Broadcasting Corporation, U.K.

E. J. ANTHONY, B.Sc., C.Eng., M.I.Mech.E., M.I.E.E.
Technical Director, Environmental Engineering, W. S. Atkins & Partners, Epsom, U.K.

J. BARNABY, P.E.
Connecticut General Life Insurance Co., Hartford, U.S.A.

M. CARATSCH, Dipl.Eng., E.T.H., S.I.A., M.A.S.H.R.A.E.
Assistant Vice President, Heating and Airconditioning Division, Sulzer Brothers, Winterthur, Switzerland.

J. M. COOLING, A.C.G.I., B.Sc.(Eng.), C.Eng., F.I.Mech.E., F.I.E.E., F.I.H.V.E.
President, Institution of Heating and Ventilating Engineers, Engineering and Commercial Director, Balfour Kilpatrick Limited, U.K.

G. P. CUNDALL, B.Sc., C.Eng., F.I.E.E., F.I.Mech.E., F.I.H.V.E., P.P.Illum.E.S., M.Cons.E. *Partner, Waterhouse & Partners, Newcastle Upon Tyne, U.K.*

D. A. DIDION, B.S.M.E., Ph.D., M.A.S.H.R.A.E.
Acting Chief, Mechanical Services Section, Center for Building Technology, National Bureau of Standards, Washington, D.C., U.S.A.

F. S. DUBIN, P.E.
President, Dubin Mindell Bloome Associates P.C., Consulting Engineers, New York, U.S.A.

D. FISK, M.A., Ph.D.
Section Head—Energy Dynamics Section, Building Research Establishment, Garston, U.K.

J. GIBSON, M.Sc., Ph.D., C.Eng., F.R.I.C., F.Inst.F.
President, The Institute of Fuel, Director, Coal Research Establishment, National Coal Board, U.K.

A. C. GRONHAUG, C.Eng., F.I.E.E.
Director of Engineering Services Development, Department of the Environment—Property Services Agency, U.K.

L. G. HADLEY, C.Eng., F.I.Mech.E., F.Inst.F., F.I.H.V.E., M.Cons.E.
Partner, Donald Smith, Seymour and Rolley, London, U.K.

J. W. HERBERT, D.L.C., C.Eng., M.I.Prod.E., A.M.B.I.M.
Manufacturing Engineering Manager, Perkins Engines Company, Peterborough, U.K.

H. JAMIESON, M.A., C.Eng., F.I.Mech.E., P.P.I.H.V.E.
Consultant, Haden Young Ltd, U.K.

H. P. JOHNSTON, D.F.H. (Hons)., C.Eng., M.I.E.E., F.I.H.V.E.
Deputy Chief Executive III, Department of the Environment—Property Services Agency, U.K.

P. J. JONAS, B.Sc.(Tech.), B.Sc.(Econ.), C.Eng., M.I.E.E., M.I.Mech.E.
Senior Principal Scientific Officer, Economics and Statistics Division, Department of Energy, U.K.

W. P. JONES, M.Sc., C.Eng., F.Inst.F., F.I.H.V.E., M.A.S.H.R.A.E.
Air Conditioning Consultant, Haden Young Ltd, U.K.

J. C. KNIGHT, C.B.E., C.Eng., F.I.Mech.E., P.P.I.H.V.E.
Consultant in Private Practice, U.K.

S. J. LEACH, B.Sc., Ph.D.
Head of Building Services & Energy Division, Building Research Establishment, Garston, U.K.

P. Le CHEMINANT B.Sc.(Econ.),
Deputy Secretary, Department of Energy, U.K.

B. LUBERT, B.Sc.(Eng.), C.Eng., F.I.Mech.E.
Chief Engineer, Marks and Spencer Ltd, U.K.

P. L. MARTIN, C.Eng., P.P.I.H.V.E., F.I.Mech.E., F.Inst.F., A.M.R.Ae.S., M.Cons.E.
Partner, Oscar Faber & Partners, U.K.

T. NICKLIN, Dip.Arch., R.I.B.A.
Partner, Ryder and Yates and Partners, Newcastle Upon Tyne, U.K.

A. F. C. SHERRATT, B.Sc., Ph.D., C.Eng., F.I.Mech.E., F.I.H.V.E., M.Inst.R.
Assistant Director and Dean of the Faculty of Architecture and Surveying, Thames Polytechnic, London, U.K.

C. P. SWAIN, B.Sc.(Eng.), C.Eng., F.I.Mech.E., F.I.H.V.E., F.I.Arb., M.Cons.E.
Principal, Colin Swain & Partners, Partner, Energy Management Consultants, U.K.

Preface

Our present civilisation is based on man's ability to harness energy and use it to his advantage. Fossil fuels have been predominant as the energy source—a concentrated form of stored energy which has been available in abundance. At last it has been generally recognised that the reserves of fossil fuel are not inexhaustible. Other sources of energy are being investigated but these are long term solutions. The immediate solution is to make better and more economical use of energy; i.e. to minimise energy use.

Fortunately the importance of energy conservation is now being realised on a world wide scale and increasingly individuals and organisations are joining the drive to save energy. Between 40 and 50% of all the energy consumed is used for controlling the environment in buildings by means of heating, lighting and air conditioning. Environmental control constitutes by far the largest single use of our available energy and is essentially the area with the greatest potential for savings to be made.

The conference on Energy Conservation and Energy Management in Buildings examined what can and is being done to minimise and economise on energy use in our present building stock, how new buildings can be designed to be low users of energy and reviewed future possibilities and prospects. A considerable amount of information was gathered together, amplified and assessed in the papers and in the active discussion. It is hoped that the conference and these proceedings will provide basic information and case studies which will be of assistance in the initiation of new energy conservation programmes.

A. F. C. SHERRATT
London W.6

ACKNOWLEDGEMENTS

The editor and technical organising committee wish to thank all the people who have participated in the arrangement and operation of the conference and in the production of these proceedings.

Contents

1

Energy Policy and Buildings

HAMISH JAMIESON

Synopsis

Buildings consume 40–50 % of the country's energy and are important when considering conservation. Heating standards have risen since 1945. Energy saving must be made in the many existing buildings, particularly in houses. Increasing insulation standards are one of many options to save energy which will be discussed. New techniques of energy control are being developed. It is not clear whether the price of fuel alone will be sufficient to produce sufficient conservation. Some important questions are set down which need to be considered as part of a National Energy Policy which should be produced by the Secretary of State for Energy who already is responsible for supplying over 60 % of the energy we need.

Sketching the Scene

THE IMPORTANCE OF BUILDINGS

I should begin by pointing out to you the importance of buildings in the field of energy use. Many of you may be aware of this already but there should be no mistake about this at this Conference. Precise figures need not trouble us but you should know that buildings absorb between 40–50 % of the primary energy used in the United Kingdom purely for creating and maintaining a comfortable internal environment. Of this, houses use about 30 % with community, commercial and industrial buildings taking another 10–20 %. So it is clear that, when considering energy conservation and energy management, there is a very good reason to take buildings as a group on their own. They are major users of energy for totally different purposes from manufacturing industry and transport.

This energy is used in buildings for heating, ventilation, air conditioning, lighting, hot water and cooking, also as power for distributing water and air and in small quantities for automatic controls, office machinery and communications. Lifts need electric power but the amount of energy they consume is relatively small.

INCREASED STANDARDS OF COMFORT

The standards of temperature which we now enjoy have increased considerably since the war. In the thirties, we used to heat offices to 15·5 °C or occasionally to 16·6 °C. Many factory buildings were then totally unheated or had a few coke stoves to take the edge off the temperature. Offices have recently crept up to 21 °C but, with other buildings, have been brought down by Government regulation to 20 °C as a maximum.

In our houses before 1939, central heating was extremely rare and we made do with a coal or gas fire in the living-room, the cooker in the kitchen and nothing anywhere else in the house, except when we were ill. And some of you will remember, with me, the chilblains we used to suffer! Now all new houses and those built over the last 20 years or so are centrally heated when built, with the result that the whole of a house can now be used throughout the winter. This has been something of a housing revolution and it is one which I think is irreversible except in conditions of crisis. However, many houses are old and their heating standards are still no better than was regarded as appropriate in the last century. A significant increase in the amount of energy used in house heating must be faced in due course, to bring these up to acceptable standards.

EXISTING BUILDINGS PREDOMINATE

Houses in total consume about twice as much energy as other types of buildings, which makes them a rich area for conservation.

There are about 19 million of them, about half in private and half in public ownership. Only a small proportion have been built in the last 25 years and the majority of them are likely to exist for many years. We need to remember in this Conference that energy conservation in buildings must be primarily concerned with existing buildings, particularly houses and this will remain the case for a long time to come. The design of new buildings, of all kinds, to low energy standards is important and will, I hope, receive proper attention here. Many of us, in fact, deal with new buildings only or predominantly and may make the mistake of forgetting that, even when built to the new standards we may agree on, they will not, in the aggregate, affect the total energy used in buildings for a long time to come.

BUILDINGS CAN USE DIFFERENT FUELS

Buildings are flexible in their use of fuel. The heating and ventilating systems in the larger domestic, commercial, industrial and community buildings can accept energy in the form of coal, oil or gas with a change only at the combustion apparatus and with the provision of fuel storage space, if required. In fact, in the last 40 years, the standard fuels used for heating have covered all of these. From the early thirties, when the growth of oil as a heating fuel was strangled at its birth by the import duty to protect the coal mines, coal or coke was the standard fuel until after the war. But from the early fifties, when the flow of oil commenced, coal was virtually abandoned and oil became the standard fuel for all but very large installations. Now, since the arrival of natural gas from the North Sea, the standard has changed again to an alternative between it and oil. A choice, where one exists between these fuels, is often not too easy to make.

Many buildings are, in fact, designed to burn one fuel to start with and with the provision of space and a suitable chimney construction, can convert to the other fuel later. While a wholesale change in the fuel supplied to buildings would impose a heavy manufacturing and installing burden of the required combustion equipment, this would be confined to new boilers or new burners and would not require different pipes or radiators. This must be an important feature in the energy strategy of the country. It is certainly one in which we, as an industry, would want to be concerned with at the proper time.

I must note in passing that electricity has now become a universal need in all buildings. It has no substitute for lighting, air and water distribution, automatic plants and television. In fact, some buildings with high lighting standards, good insulation, double-glazing and low glass areas have no other energy source but electricity. For these, our winter temperatures do not impose the need for a separate thermal energy supply.

REASONS FOR THE CONFERENCE

Why are we holding this Conference, the first certainly in the annals of the IHVE, if not of the other participating bodies?

One reason is clearly the recent and very significant rise in fuel prices. This alone would have caused us to confer on how to conserve and manage energy in buildings. That this has had an immediate effect, some of the papers of this Conference are evidence enough.

But I believe that we are now in a state of some alarm and that this is also why we are here. In spite of some lonely prophets, we lived up to the time of the Yom Kippur War as though there was as much energy in the world as we should need and in spite of our escalating requirements which might increase world consumption four times by 2000, it would never run out. We were vaguely aware that the coal reserves in this country would not last for more that one or two hundred years but we thought that, in general, there would be plenty of oil from the world at large to make up our requirements. Now, there is a wide-spread realisation that, although estimates of oil reserves can only be approximate and there may well be large resources not yet discovered, the likelihood is that thirty years hence or thereabouts, the world oil supply will begin to diminish in relation to the then world demand. And we are properly frightened at this prospect, not perhaps for ourselves but certainly for our grandchildren.

We have been astonishingly fortunate, more fortunate than some of our EEC partners, in finding fairly recently, oil and gas in the North Sea in considerable quantities. This discovery has provided an abundant supply of gas and brought this back into the forefront of choice for heating systems, when ten years ago, we were using it only in small plants. This discovery has given us a breathing space and extended the time we have to prepare for the future when the world supplies of gas and oil may be running low. We can reassure ourselves, as a result of these North Sea discoveries but only to the extent that we have a few more years to prepare for the time when the world energy crisis becomes serious.

Options for Conservation and Management

INCREASING INSULATION

Luckily, the demand for thermal energy in buildings can be significantly reduced by increasing the insulation value of their structures. Until recently, the levels of insulation used in general practice in all classes of buildings have been low, mainly because the low price of fuel did not warrant higher standards. Now we have

Regulations which, at least, insist on a much higher standard of insulation for new dwellings.

The vast problem of reducing thermal demand in existing buildings, particularly in houses, can be tackled by insulation of the cavity walls, where these exist, and of the roof spaces. The reduction of natural air change through leaky windows and doors is another fruitful and relatively inexpensive way of reducing fuel consumption. Double-glazing has advantages for comfort, even if it is still too expensive to be justified on the grounds of energy saving.

SYSTEM IMPROVEMENT AND MANAGEMENT

The heating, ventilating and air conditioning systems in community, commercial and industrial buildings offer many opportunities to save fuel. Designers can use these for new systems and building operators for systems in use. Thermostatic control can be improved and extended, heat exchangers can recoup heat from exhaust air and water. The ventilation rate may, in many cases, be reduced. Simple, good housekeeping of existing systems can be very effective, not least to run heating and air conditioning plants for as short a time as is possible. Particularly, we have been made aware quite recently by Neil Millbank of the BRE and Alec Loten of the PSA, of the waste of electric lighting in some office buildings. Here, perhaps, is the easiest way of all to save power. Ensure that the lights are turned off when the rooms are empty, even if you have to install some switching circuits to be able to do this effectively.

The Conference will deal extensively with these and other options of fuel saving and will be shown cost-effectiveness figures to indicate relative values of some.

NEED FOR KNOWLEDGE OF ENERGY CONSUMPTION

We do not know much about the total energy consumption of buildings. In the old days, when we were merely heating buildings, we normally provided an estimate of the annual fuel consumption, if only to prove that we were going to install the cheapest. In recent years, with the coming of air conditioning and the increase of lighting levels, the electric power consumption in many buildings has far outweighed the thermal energy, from a cost point of view. Estimates of the electricity consumption would clearly, in the cases referred to above, have been highly dangerous. Now we recognise the importance of the knowledge of total annual energy consumption of buildings and the

IHVE and others have begun to collect figures for energy consumption of different Building types, so the basis for some reliable knowledge of this important subject is being collated. It is probably a difficult task to collect the data accurately and to make meaningful comparisons between one building and another.

The National Bureau of Standards in Washington D.C., have, as you may hear, collected data from a number of buildings and their methods of doing this would be interesting. Energy management in buildings must have this kind of information if it is going to be able to judge its own performance as an energy user. Designers of buildings and systems likewise need it.

ENERGY BUDGETS

The 'Energy Budget', as it is called in the USA, is a new and useful factor now coming into increasing use. Broadly, this is the estimated, or actual, energy consumption of a building from all sources for internal environmental controls stated in units of energy per unit of area of building surface. It is valuable for assessing a preliminary design of a building, or for the performance of one in use. At the recent IHVE/ASHRAE meeting in Boston, this term was used quite frequently and figures for a number of buildings were given. In the United Kingdom, we have not advanced as far as this. I think we have to agree both the units we use for this ratio and also to decide whether to use the American name or coin a new one.

ASHRAE SPECIFICATION 90—P

The ASHRAE have recently produced this Specification which gives criteria for designing both domestic and office buildings for energy conservation. We heard a good deal about this in Boston and it seems likely that it will have considerable effect on the new buildings in America as it appears to be well on the way to acceptance in Federal and State Legislation. The need for something similar in this country is fairly obvious and I am glad that the IHVE are taking steps to produce a similar document.

CALCULATIONS OF COST-EFFECTIVENESS

The cost-effectiveness of energy saving measures is the basis of the economic justification for undertaking them. Whereas the capital cost of the improvements can be reasonably accurately estimated, as they would normally be the subject of tenders to carry out the work, the value of the energy saved is more difficult to calculate. This is partly because the amount of energy to be saved can only be approximately estimated and also because the cost of energy over a long period in relation to general money values is highly uncertain. Also, should the energy saving be calculated on a simple number of years to pay back the capital spent or should the calculation be based on a more sophisticated 'present value' or something similar? More important still, should not the energy costs include something to represent the value to the community of the irreplacable sources from which it comes?

Pressures to Save

WILL MOTIVATION BE ADEQUATE?

What pressures will be brought to bear on designers and on operators of buildings and systems, to conserve energy and will they be effective?

I believe that the majority of us feel that the conservation of energy in buildings is very important. We are sure that significant savings can be made and are keen to play the part required of us. But we want to have some assurance that designers and operators will be sufficiently motivated to produce results which will give them and us satisfaction. So it is reasonable to ask this question, although it has not, so far, come into our area of concern as Building Engineers very widely. Probably your presence here shows you are interested and want some kind of answer. But as Energy Conservation is largely new to us, we must inevitably speculate with the few facts available.

THE REMOVAL OF SUBSIDIES

While primary energy costs have greatly increased, the Nationalised Fuel Industries have been subsidised to hold down the cost of living. Now these subsidies are to be removed and the raising of their prices has begun. We must welcome this as an important first step in energy conservation. Social hardship, which may result, may well require alleviation. But I think it is not in our competence to comment on this.

SAVINGS BY THE HOUSEHOLDER

The higher costs of energy and those which are to come, fall directly and immediately on the householder. He is in a particularly exposed position, having to meet increases in living costs all round him. Pressure to save energy arising from its higher price will be strong. But, apart from cutting out obvious waste, can he save very much except by reducing his living standards? Probably not unless, as a house owner, he is able to spend some capital on reducing the heat loss from his house by adding insulation, on which, incidentally, he would get his money back fairly quickly.

Half the householders of the country are tenants of Local Authorities or other owners. The energy price increases will affect them too, though, in some cases, there may be some delay as charges are adjusted. Their options to save are restricted to reducing consumption and hence standards. It is their landlords who will have to spend the capital to reduce the heat loss from the dwellings.

SAVINGS BY COMMERCE AND INDUSTRY

The situation for commercial and industrial buildings is rather different. Here the fuel costs are only a part of the cost of turnover of the business that is carried on inside them. Savings in energy can be made but their significance may be judged in a different manner. The savings are likely to be but a small percentage of the cost of turnover and may be so small that they do not attract enough skilled attention from management to produce effective savings. The figures concerned are scarce but one large firm of accountants tell me that a 20 % saving on the energy charges for their offices would be about 0·4 % of their net profits before tax. For a firm of contractors, the same percentage of energy costs is about 0·25 % of net profitability. This is not to deny that energy management in such buildings cannot be effective and that the cash saved is less than its face value but the spur to achievement must arise from other motives such as a dislike of waste, not only for one's own account but also on the national level.

PUBLICITY FOR ENERGY LEVELS

As I said above, I believe that informed technical opinion must, and will be building up a corpus of knowledge about the levels of use of building energy using energy budgets and the like. Annual building energy consumptions will be brought into the foreground of discussion for the first time and will, I hope, become the subject of continuous concern at both design and management levels. Given a reasonable amount of money for research to be done into energy consumptions of existing buildings, we may hope that, reasonably soon, reliable advice on current standards can be available to designers and operators. The results from new buildings constructed to new standards of energy conservation will take longer to appear but will be important as showing the lower levels of consumption that are reasonable.

Informed technical opinion on the consumption levels, when it has been properly formed, will be a significant pressure to save, particularly if results can be made public. Something on these lines has already been published for a number of London office buildings, where the energy levels were measured and compared, though they were kept anonymous. But each building operator knew his own performance, although the public did not. Perhaps, in future, anonymity can be discarded.

GOVERNMENT LEADS TO GOOD PRACTICE

I should like to see several examples of buildings of different types specifically designed to economise in the use of all types of energy put up by the Government or Local Authorities to show what can be done. The Americans are already constructing an office block at Manchester, New Hampshire, for this purpose. Properly designed and constructed and afterwards tested, we should certainly learn something well worthwhile and, moreover, these buildings would be a mark of Government sincerity with conservation.

Energy Policy and Buildings

SOME WIDER QUESTIONS

This Conference will be mainly concerned with saving the energy used in buildings by readily available and currently understood technologies. We accept that this is the proper starting point for our industry and we are in earnest about proceeding with the work involved. But there are other questions which may or may not be discussed here, which interest us and which will need consideration and answers during the formation of a satisfactory national energy policy. Some of these are briefly stated below.

INCENTIVES TO INSULATE

To achieve any worthwhile part of the potentially large fuel saving possible in houses, will depend in the first place on whether the house owner will spend his capital in heat saving insulation. Incentives to householders in the form of subsidies are given in Holland and Sweden and the IHVE have already pressed for these to be given here also. Will these be given sometime? And equally, when will a start be made to insulate Local Authority housing?

ELECTRIC HEATING

We have got into bad habits in this country by using direct electric heaters in cold weather, mostly in houses, because the other means of heating are insufficient. This is deplorable from the point of view of the capital expenditure on and maintenance of generating plant to meet the peak loads in cold weather. The growth of electric storage heating either in the floor or by means of block heaters is a more acceptable form of electric heating bringing with it, as it does, a reasonably good load factor. This is not to say that there is no case for direct electric heating for short-term topping-up of temperature in off-peak periods or in special premises. However, all electric heating is available only after three or four times the energy used is expended in the generating station and distribution system.

Some discussion and agreement with this industry and the Electricity Council is very desirable so that we can both follow broadly the same policy when the same questions confront us.

CHOICE OF FUELS FOR HEATING

What fuels are going to be used for heating buildings in the near future? If the choice is between gas and oil, will the choice be left to each consumer, comparing the prices, availability and security of supply? Many people have questioned whether such a freedom of choice is desirable and whether the inherent advertising and competition between fuel suppliers, both of whom may soon be nationalised, should be continued. Also, both fuels have specially suitable applications and both are valuable raw materials. May the formation of a national preference for use of fuels, to put it no higher than that, restrict the availability of one of these for heating buildings?

DISTRICT HEATING

District Heating can save energy, if it burns waste or low grade fuel suitable only for large boiler plants and, particularly, if it is combined with power generation. It is widely practised in Europe. If combined with power generation, it will require a radical change in the size and location of power stations. Discussions are taking place and a number of possible schemes of combined heat and power generation are being investigated. An announcement that one or two will be proceeded with soon would be welcomed.

HOW MUCH CONSERVATION AND HOW QUICKLY?

The Building Services industry needs to be given some kind of answer to these searching questions before very long, otherwise the drive for energy conservation in buildings will atrophy and could die away to a very low level of effort. We cannot provide the answers for ourselves. Only when we have some kind of yardstick against which to measure performances will it be possible to judge whether progress is sufficient. This, in turn, will tell us whether the pressures to save energy, which up till now in this paper have been taken to arise mainly from the high price of fuel, will be strong enough.

Another question is frequently asked. Will the price of fuel alone effect sufficient conservation or will some stronger form of pressure be necessary? This is prompted, in part, I believe from the state of alarm that exists about the long-term future of our fuel supplies.

It is suggested in one of the papers you will hear, that energy budgets could be given the status of a licence, which the building owner may not exceed, or exceed only perhaps at a greatly increased cost. This is a tentative suggestion for one kind of fuel allocation, which might be necessary if the price mechanism does not work.

A NATIONAL ENERGY POLICY?

We take it that a National Energy Policy is in the process of formation and that it will cover, at least in a preliminary way to start with, the questions set out above. We hope to see the conservation programme set out against a clear statement of national energy targets. We understand the very complex nature of setting such targets and sympathise with the forecasters. We know that the data on which they must be based are pretty unreliable. It can be accepted that these will need continuous up-dating and significant changes may take place as time goes on in the main groups of figures. But both industry and the building operators require yardsticks against which to measure their performance in the continuous business of conserving energy.

A National Energy Policy also requires, in due course, a statement of the probable changes in the supplies of fuels and the sources that will be used to make good those fuels which become scarce. New technologies to provide new sources of energy are now frequently discussed and some are being researched but, so far, no policy statements about them have been made. People are already saying that buildings being constructed now will have only electricity as an energy source in 30 years time, as gas and oil will be unavailable. This is a fairly glib surmise as it takes no account of the changes in generating capacity and growth of distribution systems, which would seem to be necessary. But it does indicate current speculation which may even today have some influence on decisions.

The Secretary of State for Energy is the man in charge of the formation and execution of a National Energy Policy. We look to him to get this started in all the necessary directions and at a suitable pace. We understand that a National Policy cannot be formulated quickly but must emerge over some years. The present policy of the Government of exhortation to 'Save It' is a reasonable beginning but very much a beginning. We would like to hear something soon of the next steps.

The Secretary of State is also in charge of the Nationalised Fuel Industries of coal, gas and electricity generation and, thus, is responsible for the supply of over 60% of our primary energy. Through the British National Oil Company, he will be seeking to supply part of the remaining primary energy which comes from oil. Such a concentration of powers of both planning and supply is perhaps an advantage. Responsibility is clearly undivided. In the light of the growing importance of energy in our national life and the disasters to it a future restriction of supply could bring, the office of the Secretary of State for Energy must be regarded as potentially one of the most important in Government.

Discussion

J. E. Greatorex (*J. E. Greatorex & Partners*): Thirty years ago district heating was considered economical at housing densities of about 12 to the acre and yet we are only now beginning to take district heating seriously.

I would like to raise the question of pricing energy from the point of view of the layman consumer. The bills come in at say, 2p a unit for electricity, 12p a therm for gas, £40 per ton for coal. How can they compare the cost? They cannot. The difficulties have been made clear to us recently in investigations relating to local authority housing with direct electric heating. I would suggest the adoption of a standard unit of energy to enable energy costs to be easily assessed and compared by the average domestic consumer.

W. J. Ablett (*Ministry of Defence*): Mr. Jamieson makes reference to the need for the Government to produce examples of buildings of different types, specifically designed to economise in the use of all types of energy.

I would point out that some buildings have been designed with this in mind. In the early 1960s, research in the Department of Health produced evidence that the total revenue costs for a new hospital were approximately equal to the capital costs after about 4 years. This led to the development of the 'Best-buy' hospital concept, and two similar hospitals were designed and built, using 'present worth' techniques with costs taken over 60 years. The height, shape, orientation, internal colour finishes, internal traffic movement, lifts, fenestration (30 %), 4 roof-top boiler houses with interconnecting ring distribution, natural ventilation, natural lighting and thermal structure (which incidentally, meets the new 1975 Building Regulations) were all taken into account. The capital and revenue costs were lower than any other comparatively sized hospitals. A number of the ideas incorporated could be utilised elsewhere. Many visitors came from other countries to see the results of this new technique.

Secondly, reference is made to the utilisation of waste heat and district heating associated with power stations. Again in 1960, a military power station was designed incorporating this, so the Government have provided an example.

A particular point on new developments is that with the implication of the free structure. We have a situation whereby if we employ consultants, their fees are based upon the capital costs of the project and are not related to revenue costs. I am beginning to think that there is a case for introducing a new form of fee structure for Architects, Engineers and others which takes into account the revenue consequences. Perhaps the fee should be inversely proportional to the revenue costs. Consultants should certainly be more closely associated with the subsequent problems arising from their designs—at least for a period of three years after it becomes operational.

J. Peach (*Institution of Heating and Ventilating Engineers*): Mr Greatorex and Mr. Jamieson have both mentioned the question of units, a subject in which I have had more than a little interest over the years. There is a perfectly good unit for energy and that is the joule and its various multiples, all part of the International System (SI). The only trouble is that the joule is small in magnitude and nearly always requires prefixes.

Due, no doubt, to various pressures internationally over the last few years, two further prefixes were agreed at the international meeting of the General Conference of Weights and Measures (CGPM) in June 1975. One of these is peta for the multiple 10^{15} and takes the symbol P and the other, exa taking the symbol E for the multiple 10^{18}.

As far as energy use is concerned this means that in general terms Domestic Dwelling Annual Energy Consumption would be conveniently expressed in gigajoules; Factory Consumptions in terajoules; Industry-wide Consumption in petajoules and National Consumptions in exajoules.

W. R. H. Orchard (Orchard Partners): There have been references to the comparison between electrical costs and heating of domestic dwellings, a building type which is responsible for a large proportion of the national heat consumption. Alternative fuels and district heating have been mentioned. The Government should seriously consider reviewing the allowance for district heating in Local Authority housing which is pegged currently at £340 per dwelling, even in cases where higher capital figures are justified based on an economic analysis.

If we examine generating capacity required for an electrically heated dwelling and take this as being approximately 7 kilowatts, which is certainly conservative, to cover the heating and hot water load, and then look at the cost per kilowatt for a new electrical generating station the figures are staggering. I understand that the cheapest type of new generating station, which would be a conventional plant, costs approximately £120 per kW. This is equivalent to the capital expenditure nationally of £840 on additional generating plant for every extra electrically heated dwelling. The proposal has been made that we will soon have all nuclear stations. The present day cost for these stations, not taking into account either distribution or running costs, is £340 per kW. This amounts to £2380 per dwelling. (1974 figures from Electricity Council UK.)

One can do a phenomenal 'Rolls Royce' of a district heating scheme for £2380 per dwelling. In fact, it would be difficult for any consultant to spend that amount of money. It has been suggested that in 30 years we will have an all electric energy economy. I would ask what will happen to waste heat which must inevitably be rejected on account of the basic thermodynamic cycles which have to be used. Surely today we should be calling for a national hot water grid—of the kind being seriously considered in a number of European countries. Unless we start now to build a national hot water grid which must start with small local stations in exactly the same way as our gas and electrical industries originally started, then when the time comes we will find we have no mechanism for distributing low grade heat, and will have wasted years of our national resources putting in individual appliances, whether gas or electric, which will themselves have consumed our scarce resources at a faster rate than would have occurred had we had district heating and a national hot water grid. It would seem that to develop a national hot water grid we require a national energy policy where all the above factors can be taken into account so that in 50 years time waste energy can be distributed and could be inexpensive.

H. C. Jamieson: I did not exactly say all buildings in 30 years' time would be heated electrically but that some people were thinking they might be. I support your general comment. I do not think there is sufficient evidence that the subject is having the discussion in the Department of Energy it deserves.

A committee has been set up to consider the restrictions imposed by the Electricity Act on the sale of heat. I am glad to know the committee is doing this but also hope it may be given powers to act and to start one or two pilot plants on a small scale scattered around the country. The last report of the Electricity Council mentioned that up to 20 schemes for combined heat and power generation were being looked at. We would be very much more interested if we heard they were going to start something.

J. M. Cooling (Balfour Kilpatrick Ltd. President—Institution of Heating and Ventilating Engineers): The scheme to which Mr. Ablett referred was, I think, at Aldershot and has been operating for some fifteen years, Are the operating figures available or is it secret because it is defence.

W. J. Ablett: Annual Reports were being produced until 1971. They were temporarily discontinued, but the Property Services Agency of the Department of the Environment have produced Trading Accounts and could probably make some of this information known to interested parties.

G. Haslett (Electricity Council): May I remind Mr. Orchard that most of the electric heating installed is storage and that there is not a great deal of capital contribution involved.

W. R. H. Orchard. (In reply to Mr. Haslett): You have only to look at the Electricity Generating Board's figures for peak Summer and Winter loads at a weekend period to see the effect that electric heating loads have on the generating peak. In addition the hot water load in domestic dwellings does not vary so significantly between Summer and Winter. I do not want to go into great detail but the last statement in relationship to night storage heaters and the fact that they only operate off peak is not strictly correct, particularly as there has been strong recent promotion of direct types of electric heating, either ceiling heating or panel radiator heating.

Extension of off peak to White Meter and Peak Boost has added further loads. Also the diversity factor on night storage and off peak equipment proved embarrassing, we gather, due to the fact that all the units switched on for charging at the same time.

M. Swiss (Freelance Journalist): We are very much aware that whatever is planned or suggested today cannot be accomplished for twenty years. That is the time lag inherent in almost all large scale engineering. To proceed properly a non-political forecast is needed which could be based on a census of energy consumption. It should be possible, with computerised techniques, to obtain figures for energy consumption in a variety of forms for all households and industry.

It is a common misconception that district heating is feasible on the continent in countries like France and Sweden because buildings stand closer together, in fact, half the French population live in isolated houses! It is also perhaps surprising that about a quarter of the housing in France was built before 1914. Therefore, when we project for thirty years ahead, predominantly it will be today's buildings that are going to house the users of electricity, gas and any other form of heat thirty years from now.

National heating grids have been mentioned—Germany is now carrying out investigations for a national heating system to be completed fifty years from now and Soviet Russia is carrying out schemes which, although no specific date is given for them, would connect areas as distant as Vladivostok and Moscow by hot water pipelines. These things have to be looked at realistically and given the right time scale.

A. W. Brown (*Clorius Meters Ltd*): We have had figures and statistics given to us, one of which showed the world requirements for energy will quadruple by the year 2000. We also learn that at the same time most of our natural energy resources are going to run out. This is an impossible situation the effects of which are quite horrifying. Is it the intention of this conference to give us a breathing space by setting conservation targets and suggesting some positive incentives? These should be of a realistic nature, not 10%, giving us only a couple of years, not even 20%, but aiming for 50% savings so that we really have a little longer to develop alternative methods of providing heat, suitable ways of distributing it and an established discipline for controlling and monitoring its use.

H. C. Jamieson: I do not think the conference could go quite as far as you would like. I made the point that we need nationally set targets and a national monitoring system to enable us to see where, as a nation, we are going in relation to energy consumption and supply. This conference will not do it, but it could point the way and thereafter it is a matter for the Department of Energy. At the IHVE Meeting in Boston in June 1975, I was very impressed by the two speakers concerned with US energy policy. Both of them came out very strongly for national targets.

To respond to Mr. Ablett, the Ministry of Defence's combined power and heat schemes at Aldershot is a shining example of good practice. I should like to see this kind of installation repeated in other groups of buildings.

The pricing of a hospital on a 60 year life cycle is also very interesting. The Americans have done something similar. They have a project for building offices priced on a 40 year life cycle. The tender price includes capital cost plus 40 year running costs. If you save one dollar per annum on running costs, it equals 40 dollars on the capital costs.

In the broad, as Mr Swiss said, there are very long lead times on energy saving projects on a national scale. This is what worries me. The Department of Energy will have to make a number of clear technical decisions in some of the areas I have mentioned. They have to make them because nobody else can and there is not much time to do it.

2

Legislation and Energy Conservation

C. P. SWAIN

Synopsis

This paper is concerned with the practical rather than the legal aspects of legislation as it affects energy conservation in buildings. The more important current legislation is reviewed, with a critical analysis of its application to present day needs. Consideration is given to the practicability of introducing direct statutory control of energy use. Suggestions are made for the desirable pattern of future legislation in this field.

Introduction

At least 40 Acts of Parliament affect energy conservation in one way or another, as do numerous Regulations, Orders, and other statutory instruments. Many more Explanatory Memoranda, Departmental Circulars and the like carry some official authority although they lack legal force.

Most of this legislation was prepared primarily for purposes quite different from energy saving and its influence on energy is merely an indirect by-product. For example, only one Act of Parliament has fuel economy as its declared objective.

Discussion of present legislation is confined to that applicable in England and Wales, and differences in Scottish Law (such as in Building Regulations) are not considered. The viewpoint is that of a practising engineer rather than that of a lawyer.

It is axiomatic that energy is a valuable and diminishing resource which must not be squandered by wasteful use, but that energy conservation has to be treated with a sense of proportion rather than by imposing indiscriminate restrictions on environmental or industrial standards. It is in this context that legislation affecting energy conservation is considered.

Present Legislation

Statutory controls, particularly those of a technical nature, are usually covered by subordinate rather than primary legislation. The Act of Parliament takes the form of a basic framework that sets out the general policy and gives the appropriate Minister the power to introduce regulations and orders for specified purposes. It is these statutory instruments which constitute the practical working part of the legislation—the Building Regulations being a well-known example. The subordinate legislation prepared under an enabling Act must be laid before Parliament, but is not necessarily debated.

This procedure permits a Minister to proceed with technically complex proposals on advice by experts within and possibly outside his Department, and without the scrutiny of a Parliament that cannot be expected to appreciate the finer points. Although it brings the risk of subordinate legislation being badly thought out or poorly drafted, the use of statutory instruments is probably the most appropriate form of legislation for the type of measure with which we are concerned.

Let us examine some of the legislation that affects the utilisation of energy in buildings.

ENVIRONMENTAL STANDARDS

The various statutory standards for temperature, lighting, ventilation, etc. (*e.g.*[1–3]) are mainly minimal requirements for health, safety and comfort which no-one would suggest should be relaxed. Indeed, measures for energy conservation should be so devised as to permit appropriate improvements to such standards. But regulations setting minimum standards of daylight should perhaps be reviewed in present circumstances, as these can lead to excessive fenestration with thermal discomfort both summer and winter and disproportionate energy consumption.

These minimum statutory standards have been in force for some time. Only recently, with the Order hastily introduced last year,[4] has there been energy-saving legislation which stipulated maximum environmental standards. This Order limited the temperature maintained by heating systems in commercial and industrial premises to 20 °C. But in recognition of the many substandard installations with inadequate flexibility or control, the Order expressly excluded cases where compliance would cause the temperature elsewhere in the premises to fall below the statutory minimum. So the Order effectively applies only to properly balanced and controlled systems that are those most likely to be reasonably efficient, while inflexible systems that are probably the very ones most wasteful in their energy consumption are exempted. This is an effective illustration of the anomalies that can arise with prescriptive legislation for one parameter in isolation. Such legislation is of limited efficacy and may be unfair in its operation.

INSULATION OF FACTORIES

Standards were established in 1958[5] for the insulation of factory roofs, prescribing a maximum U value (excluding glazing) of $1{\cdot}7\ \text{W/m}^2\text{K}$. The requirement was deemed to be satisfied if the heat loss through any square metre of roof did not exceed 38 W with an outside temperature of -1 °C and when the specified internal temperature was measured at a level of 1·5 m above the floor.

At the time of their introduction these standards were criticised for being too low, for failing to specify an insulation standard for walls, and for taking no account of temperature gradients. But this was a new type of legislation, and at the time the Regulations were doubtless a reasonable compromise between the various conflicting considerations. However the time is now opportune for amendments and strengthening of these Regulations to suit present circumstances.

A feature of these Regulations that appears all too often in technical legislation is a principal requirement that can be clearly understood and easily applied, coupled with a 'deemed to satisfy' clause that is more obscure rather than simpler. In this instance the 'deemed to satisfy' clause also changes the meaning of the primary requirement as it permits a higher transmittance if the inside design temperature is less than 21 °C.

INSULATION OF DWELLINGS

Standards of insulation for dwellings are covered by Building Regulations, which are issued under powers derived from the Public Health Acts of 1936 and 1961. These powers were limited to Regulations affecting health and did not extend to measures that might be desirable for reasons of energy conservation. So the reason for the higher standards of thermal insulation required by the 1974 Amendments to Section F of the Building Regulations[6] is to prevent condensation and mould, and the consequent saving of energy is merely an ancillary benefit. Any Regulations introduced for the primary purpose of saving energy would have exceeded the Minister's powers and been ultra vires. The position has since been changed by the Health and Safety at Work Act 1974, which enables the Minister to make Regulations for 'furthering the conservation of fuel and power', so the scope for introducing commonsense energy-saving legislation is now less restricted.

Before the recent amendments the requirements for insulation of dwellings[7] took the form of limiting U values for each external element (except glazing), with no restriction on the amount of glazing or of the heat loss from the structure as a whole. So these earlier Regulations, although easy to understand and apply, did little to assist energy conservation. An IHVE Report[8] recommended a drastic reduction in the permitted elemental transmittances coupled with a limitation of the heat loss per unit volume of the dwelling. These proposals were extended in a paper by Billington,[9] with further suggestions for distinguishing between light and heavy structures and taking account of climatic zone.

TABLE 1
COMPARISON OF INSULATION STANDARDS

	Building Regulations 1972	*Building Regulations* 1974	*IHVE Report*	*Billington Paper*
Max. U value W/m^2K				
Roofs	1·42	0·6	0·6	0·4–0·6
External walls (excl. openings)	1·7	1·0	0·6	0·4–0·6
Party walls*		1·7		
Exposed floors	1·42	1·0	0·6	0·4–0·6
Average of perimeter walling (incl. glazing)		1·8		
Max. fabric loss per unit volume of building W/m^3K			0·9	0·5–0·9

Billington proposes different values according to:
(a) the mass of the structure (2 categories)
(b) the climatic zone (3 categories)
* In the 1974 Building Regulations, the U value of party walls to other dwellings also subject to the Regulations is deemed to be 0·5 W/m^2K.

TABLE 2
TYPICAL HOUSES

	Building Regulations 1972	*Building Regulations* 1974	*IHVE Proposals*
Detached House			
Thermal transmittance W/m²K			
External walls	1·7	1·0	0·6
Roof	1·4	0·6	0·6
Floor	0·45*	0·45*	0·45*
Window area m²	28*	23	14
Percentage glazing	20*	17	10
Fabric heat loss W/°C	440	301	207
Specific fabric loss W/m³K	1·9	1·3	0·9
Average U value for perimeter walling W/m²K	2·5	1·8	1·1
Total heat loss kW	9·8	7·2	5·4
Semi-detached House			
Thermal transmittance W/m²K			
External walls	1·7	1·0	0·6
Party walls		0·5	0·5*
Roof	1·4	0·6	0·6
Floor	0·35*	0·35*	0·35*
Window area m²	28*	28	20
Percentage glazing	30*	30	22
Fabric heat loss W/°C	357	273	207
Specific fabric loss W/m³K	1·6	1·2	0·9
Average U value for perimeter walling W/m²K	2·1	1·8	1·3
Total heat loss kW	8·2	6·6	5·4

* Indicates assumed value.

In the new Regulations[6] that came into effect earlier this year the permissible U values are reduced, albeit not to the extent advocated by Billington. There is now a limitation on the overall fabric heat loss and hence on the proportion of glazing. However this is by restricting the average U value of the perimeter walling rather than a volumetric basis, and this distinction has important consequences.

The significant features of the previous and current Building Regulations are shown in Table 1, and are compared with the recommendations of the IHVE Report and of Billington's paper.

Table 2 shows the practical effect of the new Regulations on typical detached and semi-detached houses of 100 m² floor area and compares the alternative approach recommended in the IHVE Report.

Table 3 is for detached bungalows of the same floor area as the houses in Table 2. One bungalow is of compact rectangular plan while the other is an irregular shape with 30% more external wall. The current Building Regulations are compared with the alternative proposals of the IHVE Report.

TABLE 3
DETACHED BUNGALOWS 100 m^2

	Compact Plan		*Distended Plan*	
	Building Regulations 1974	*IHVE Proposal*	*Building Regulations* 1974	*IHVE Proposal*
Transmittance W/m^{2}K				
External walls	1·0	0·6	1·0	0·6
Roofs	0·6	0·6	0·6	0·6
Floor	0·35*	0·35*	0·35*	0·35*
Window area m^2	16	11	22	7
Percentage glazing	17	12	17	5
Fabric loss W/°C	264	207	327	207
Specific fabric loss W/m^{3}K	1·2	0·9	1·4	0·9
Average U value for perimeter walling W/m^{2}K	1·8	1·2	1·8	0·9
Total heat loss kW	6·5	5·4	7·7	5·4

* Indicates assumed values.

Tables 2 and 3 are for dwellings designed to the limits of the relevant requirements. All windows are taken to be single glazed. Impracticably small window areas calculated on this basis indicate the need for a higher standard of thermal insulation and/or double glazing to enable larger windows to be provided.

The notional nett annual energy requirements of various types of dwelling are shown in Table 4. The heat losses are calculated on the same basis as for Tables 2 and 3, and it is assumed in each case that the calculated gross energy requirements are offset by fortuitous internal gains equivalent to 15 GJ/yr.

Examination of these stylised comparisons leads to the following conclusions.

(1) The current Building Regulations do not ensure adequate thermal standards in the design of new dwellings for the optimum utilisation of energy.

(2) A reduced standard of insulation is demanded for non-detached dwellings—flats and terraced or semi-detached houses—than for the corresponding detached dwellings. It seems equitable that the non-detached dwelling, having a smaller proportion of external walling, should be permitted larger windows than its detached counterpart.

(3) Building Regulations in their present form actively discourage an efficient shape of building plan, as straggling buildings are permitted to have larger window areas than compact ones. A standard based on maximum heat loss per unit volume or unit floor area would penalise inefficient shapes by requiring better insulation.

(4) Standards of the form proposed by IHVE and Billington take no account of differences in size and require smaller dwellings to be better insulated than larger ones. This may be logical from economic considerations, but is surely unacceptable on social grounds.

TABLE 4
NOTIONAL NETT ENERGY REQUIREMENTS, GJ/ANNUM

	Floor area m^2	*Building Regulations* 1974	*IHVE Proposal*
Intermediate flat	100	27	39
Detached bungalow, compact	100	50	39
Detached bungalow, extended	100	62	39
Semi-detached house	100	52	39
Detached house, compact plan	100	57	39
Detached house, compact plan	200	98	94
Detached house, extended plan	200	121	94

The present statutory insulation requirements for dwellings should, therefore, be considered as purely temporary expedients. New standards are needed which:

(a) demand better insulation of the fabric elements than do the present standards;
(b) impose overall restrictions on total heat loss per unit volume or unit floor area;
(c) take thermal capacity of the fabric into account;
(d) provide different requirements according to local climate;
(e) allow for differences in size of dwellings.

DISTRICT HEATING AND THE ELECTRICITY ACTS

Electricity supply is governed by a dozen Acts of Parliament, of which the most important as regards district heating are those of 1919 and 1947.

Before nationalisation electricity was supplied by some 500 private electricity companies and local authorities. Under the 1919 Act these authorities had a statutory duty '... to provide ... a cheap and abundant supply of electricity within their district'. But energy conservation was in mind even at that time, as the Act also empowered supply authorities to use waste heat for power generation and to supply '... any form of energy other than electricity'.

On nationalisation in 1947 the provisions concerning district heating changed. The clause permitting the supply of other forms of energy was replaced by a seemingly stronger provision which was mandatory instead of permissive. The present statute now reads 'It shall be the duty of every Area Board to investigate methods by which heat obtained from or in connection with the generation of electricity may be used for the heating of buildings in neighbouring localities, or for any other purpose, and the Area Board may accordingly conduct, or assist others in conducting, research into any matters relating to such methods of using heat. Any Area Board may themselves provide, or assist other persons to provide, for the heating of buildings by such methods ...'.

Although the wording of this clause is by legislative standards a model of clarity, it is doubtful whether the practical implications were sufficiently considered. Under the various Acts, Electricity Boards have statutory duties:

(1) to provide a cheap and abundant supply of electricity (Section 8(1) of the 1919 Act);
(2) to promote the use of all economical methods of generating, transmitting, and distributing electricity (Section 1(6)(a) of the 1947 Act);
(3) to operate tariffs that do not discriminate for or against any class of user (Section 37(8) of the 1947 Act);
(4) to investigate methods of supplying heat for district heating schemes (Section 50 of the 1947 Act).

Whereas the first three obligations require Area Boards to take positive action, the obligation concerning district heating is for investigation, and implementation is purely permissive. It is presumably a matter of opinion as to how thoroughly such investigations must be conducted.

The economics of combined thermo-electric generation are complex and far from clear-cut, as is shown by numerous technical papers with conflicting conclusions (*e.g.*[10, 11, 12]). The only certainty is that the efficiency of electricity generation is reduced, regardless of any gain from the combined operation. The prime obligation of the Electricity Boards remains the provision of cheap, abundant, and economical supplies of electricity without tariff discrimination. Their statutory duty concerning district heating is discharged once such schemes have been investigated. Any Board implementing a district heating scheme lays itself open to a charge of breach of statutory duty being brought by any disgruntled electricity consumer, and would have to produce irrefutable evidence to show that electricity consumers were in no way subsidising thermal users. It is not surprising that the electricity supply industry has with few exceptions complied with its statutory duties but not laid itself open to the difficulties and troubles that could follow the sale of heat. It is not suggested that this is the only or even the main reason for the rarity of combined thermo-electric generating schemes in this country. But it is contended that the particular wording of the 1947 Act encouraged the pursuit of ever-increasing efficiency of electricity generation to the detriment of studying the optimum utilisation of basic fuels. Although the Department of Energy is now encouraging a re-examination of the electricity supply industry's proper role in district heating, any new thinking will be fruitless if confined by the present statutory obligations.

FINANCIAL LEGISLATION

The Government can clearly influence energy usage by its policies on taxation, subsidies, pricing directives, and the like. Such measures should be consistent with the overall energy strategy, but detailed consideration is a matter for economists and is outside the scope of this paper.

The Government can also influence energy conservation at domestic, commercial, and industrial levels both positively by incentives and negatively by restrictions. Recent examples are the extension of full capital allowances to thermal insulation in the April Budget, and the imposition last year of credit restrictions on domestic central heating which were later lifted after representations about the adverse effect on fuel economy.

But Governmental measures that are or appear to be half-hearted can do more harm than good. When the Energy Saving Loan Scheme for Industry was announced last autumn it seemed that the Government was at last beginning to take energy conservation seriously. But it transpired that the Scheme applied only to projects costing more than £10 000 which would repay capital in three years, and that commercial rates of interest would be charged (initially 13½%, later reduced to 11½%). How much more credible would have been the Government's exhortations to save energy if they had not permitted the Treasury to have the last word but had offered more realistic loans at half the current commercial rates of interest. It is difficult to believe that the improvement in the balance of payments would not have justified the extra internal expenditure.

The Government's declared policy is against providing fiscal incentives for energy-saving measures as '... there is no general case for the State bearing the cost of cutting fuel bills'.[13] This attitude is, in the author's submission, quite inconsistent with the real needs of the present time and is indicative of a misplaced order of priority. Is energy conservation less worthy of support for reasons of national interest than Government aids for home improvements, industrial development in depressed areas, building hotels—or indeed than housing, rates support, or food subsidies?

IMPLEMENTATION OF REGULATIONS

Statutory instruments such as the new insulation standards in Section F of the Building Regulations must be administered by Local Authority inspectors and implemented by architects, engineers and surveyors—not all of whom will be readily conversant with calculations of U values and heat losses. Practical application of the 1974 amendments is made more difficult than it need be by the presentation and legal jargon of the 'deemed to satisfy' provisions. The mass of verbiage and tedious cross-referencing makes the requirements almost incomprehensible until translated into plain English. If before these amendments came into force most local authority building control officers had produced their own simplified formulae,[14] why could not the amendments themselves been made more readily usable in the first place?

The obscurities occur mainly in the 'deemed to satisfy' provisions, which take the form of listed types of construction that are held to comply with the requirements without specific proof. This, it is submitted, is an unwise use of 'deemed to satisfy' provisions as it tends to inhibit the exploration of new forms of construction whose compliance with the Regulations has to be demonstrated by calculation. There can also be conflict with other parts of the Building Regulations, as happened with cavity fill insulation. This form of insulation is deemed to satisfy Section F but contravenes Regulation C9 (which prohibits bridging a cavity wall). So Local Authorities had to be advised by the Department of the Environment to relax requirement C9 in appropriate circumstances.[15]

Section F of the Building Regulations is concerned only with U values, so it would appear sensible for the Regulations to stipulate the method by which the U values are to be determined. But this is not done, and it is left to a non-mandatory Information Note[16] to recommend that the procedure of IHVE Guide Book A should be followed. This shows a misplaced order of priority. Surely it would be better for the Regulations to stipulate the method of calculation that is 'deemed to satisfy', and for the Information Note to include types of construction that should be considered to satisfy the statutory requirements.

Controlling Energy Consumption

Future legislation must be directed towards conserving energy on a national scale by ensuring that it is efficiently utilised. This requires some form of control over every individual user. Energy-saving measures must be realistic and practicable. Any statutory requirement that necessitated the consumption of more energy to administer it than the amount of energy saved would clearly be self-defeating. Any legislation for thermal insulation that stimulated a demand beyond the resources of the insulation industry would be worthless.

In a free economy it is desirable that energy-saving legislation should be persuasive rather than directive. But exhortation alone cannot be an effective long-term means of energy economy, and the benefits of voluntary savings promoted by the prolonged advertising campaign of the Department of Energy must diminish as the immediate impact of the energy crisis passes. So other means of controlling energy consumption must be considered. In this consideration the differing effects on commerce, industry, and the private household must be recognised.

CONTROL BY PRICING

Few would disagree that energy prices have been too low in the past and must be set at more realistic levels in the future—unpleasant though the transition is. But the realistic pricing of energy is not sufficient in itself to promote efficient utilisation.

In much of industry and commerce the cost of energy is only a fraction of production and labour costs, and the incentive to economise is small. Effort and expenditure is more profitably directed towards improving productivity or reducing transport costs than to saving energy in buildings. This does not apply to the larger organisations using massive quantities of energy, where the employment of in-house expertise is demonstrably cost effective. It is the multiplicity of small firms with lower energy demands and less efficient utilisation that are the least likely to economise no matter how high the cost of energy rises.

On the other hand energy costs are an important proportion of the householder's budget, and in the long term higher prices will result in greater economy. But this economy can only be achieved by better insulation, better controls, better heating systems and better housing. Many house owners cannot afford such capital expenditure, particularly at the present time of rapid inflation, and should be assisted with grants or cheap loans. But this would do nothing to assist tenants of rented property, and it is the poorer and less able sectors of the community who suffer most from high domestic tariffs. The need is for efficiency not deprivation, and high energy costs will fall most heavily on those least able to bear them.

FINANCIAL INCENTIVES

The case has already been argued for incentives in the form of a realistic loan scheme for industry, tax concessions, and grants for householders. Such means for encouraging expenditure on energy-saving measures are unlikely to be sufficient in themselves and must be supplementary to other forms of control.

PRESCRIPTIVE CONTROLS

Energy consumption is determined by so many inter-related factors that even if desirable it is not practicable to legislate for all but a few. It would be possible to set statutory standards for temperature, lighting, insulation, and boiler efficiency. But one cannot ensure by legislation that heating or air-conditioning plant is properly selected for its function; that distribution systems are balanced; that appropriate thermostatic controls are both fitted and intelligently set; that lights are switched off when not needed; or that heat is not wasted through open doors and windows.

Prescriptive regulation of individual design and operational factors is not only of limited effectiveness in controlling energy usage, but it also stifles design initiative. The greater the extent of control the more difficult are the regulations to apply and administer, and the greater is the probability of anomalies and unforeseen snags.

DIRECT CONTROLS

Indirect methods of controlling energy consumption are either impracticable or at best highly imperfect, so serious thought must be given to the direct control of energy use. This is not synonymous with energy rationing since energy has no intrinsic value and the restriction is of energy waste rather than its beneficial products.

Any form of direct energy control must eventually apply to all buildings without exception and be directed primarily towards relatively small users. Households and small firms account for half our energy consumption in buildings and are collectively the least efficient in its utilisation. But the problems of introducing a system of direct control would be enormous and implementation would have to be in stages. Perhaps the simplest to tackle are commercial buildings, as there is little variety in their function and almost all their energy is used for building services. Industrial buildings are more difficult because of their greater range of function and their considerable use of energy for machinery and process purposes. The application of energy consumption controls to dwellings would cause obvious practical and social problems, but this is the least urgent sector as it is the one where indirect controls are likely to be the most effective.

Various proposals have been made for methods of directly controlling energy use (*e.g.*[17, 18]), which have been freely drawn upon in setting out the energy licencing scheme that is advocated here. These methods of direct control inherently rely on establishing a consumption norm against which actual energy use is measured. Direct controls of this type are criticised on grounds of inflexibility, clumsiness, expense of administration, and inappropriate State intervention. Such criticisms cannot be dismissed lightly, but a decision has to be made between various conflicting desiderata. Which is the more important—energy conservation or individual freedom of choice? It is submitted that the direct control of energy consumption is a necessary evil, and that a scheme of energy licencing would be both practicable and effective.

Proposed Energy Licensing Scheme

To control energy consumption, everyone using energy in a non-domestic building must have a licence which will stipulate the maximum amount of energy that may be purchased each year for all purposes except transport. Norms will be established for the reasonable energy demands to enable satisfactory conditions to be maintained in buildings of moderately good thermal characteristics. The limit for each consumer will be based on these norms and assessed on the floor area, location, and use of the building. Users will be free to choose the forms of energy they purchase, but the limitation will be on gross energy value (including energy used for production and distribution) so the nett energy content will be weighted to deter the

TABLE 5
ENERGY WEIGHTING FACTORS

	Relative gross/nett energy ratio	*Suggested weighting factor*
Coal	1	1
Natural gas	1·04	1
Oil	1·07	1·2
Manufactured fuels	1·37	1·4
Electricity	3·65	3·5
Self-generated (solar, wind, tidal, geothermal, etc.)		0·5

improper use of high grade energy. Table 5 shows the ratio obtained by strict application of this criterion[19] together with suggested practical values that allow for other relevant factors.

The norms must be reasonably generous, and will be designed to prevent extravagent use rather than to impose harsh penalties on those struggling for better efficiency. To discourage efficient users from increasing their consumption to the permitted limit there must be incentives—perhaps in the form of tax concessions—if consumption is substantially below the allowance. Conversely, consumers with plant that cannot be made to conform to the required limits except at disproportionate cost must be given special consideration.

With the energy licencing scheme it would not be necessary to restrict healthy competition between fuels, and advertising campaigns designed to sell more energy would die a natural death. The Trade Descriptions Act 1968 already gives protection against inaccurate advertising claims, although it is probably not implemented often enough in the energy field. Tariffs offering reduced rates for larger consumption could remain as they stimulate efficient distribution, although without any overall restriction they also encourage extravagance.

A requirement of Building Regulations consent for all new buildings must be a demonstration that the energy usage can be within the licenced limit. In the first few years of the scheme any consumer unable to comply will have to seek a supplementary licence, which will permit additional energy to be purchased but at progressively increasing penal rates to encourage upgrading of older plant.

Establishing energy norms for commercial buildings will be relatively straightforward. There must be separate norms for offices, shops, hotels, restaurants, and warehouses which take account of building size and location, the extent if any of air-conditioning, and any special factors such as computer rooms or abnormal working hours.

Energy norms for factories can either be on an industry basis by assessing the energy content per unit of production, or alternatively by energy allowances for each process and machine type. A rational system will clearly be difficult to establish, and initially other control measures must be applied to industry. The most appropriate would be a compulsory energy audit whereby every company is compelled to include with its annual accounts an energy balance showing the amount and manner of its energy use during the year. This must be in an approved form so that cases warranting further investigation can readily be seen.

The process of monitoring energy consumption can be simple. The number of energy suppliers is small and most use computerised accounting systems so copy accounts can easily be passed to a central licence administration office for processing. The administrative costs will be low, and could be recouped either by charging for licences or by a surcharge on energy prices. The costs of administering the scheme need not increase fuel costs by more than about one half per cent. The heavier initial expenses of setting up the scheme, establishing appropriate norms, and setting licenced limits could be covered by the extra charge made for energy supplied against supplementary licences.

It is suggested that the licencing scheme is administered on a national basis by the Department of Energy, who will establish energy norms and standard procedures for applications and appeal. A register of suitably qualified technologists in private practice should be formed with authority to certify appropriate energy limits for each licence application. This would obviate the need for large staff increases by Local Authorities. There are precedents for the statutory recognition of panels of independent experts,[20] and in this case their role would be analogous to that of an accountant. To pursue the parallel, these approved experts might also audit the energy accounts that are suggested for industry.

Indirect Controls

The imposition of direct controls on energy consumption would reduce the need for prescriptive controls on individual design parameters. But it would be unwise to relinquish indirect controls until such time as they have outlived their value.

New standards should be introduced for the insulation of dwellings based purely on energy-saving considerations and in the form previously advocated. Size and shape become more significant with commercial buildings, and new regulations should be designed to penalise buildings of extravagent shape irrespective of size. Nicholson and Fitzgerald have considered this point[21] and suggest a limitation on the permissible mean thermal transmittance that is related to a reasonable value for a building of optimum shape. The value for the optimal building would be adjusted in the ratio of the building volume to the power two thirds divided by the exposed surface area to obtain the maximum permitted mean U value for a particular building. Such limitations would not be appropriate for industrial buildings, where simple maximum transmittances for roofs and walls would suffice, provided they are coupled with restrictions on glazing.

Prescriptive limits for temperature, lighting and the like are, it is suggested, unnecessary. It would be convenient to have minimum standards of plant performance and response, but these are not suitable subjects for legislation.

Future Legislation

Future legislation must be enacted within the framework of a national energy policy that has continuity and is not allowed to become a football for party politics. This consistency must apply to any Government measure that has a potential effect on energy. Legislation concerning energy must be both self-consistent and more readily comprehensible than in the past.

The objects of future legislation should be to provide the necessary incentives and controls to promote efficient energy utilisation, to remove impediments such as those that effect the implementation of district heating schemes, and to correlate the activities of the energy industries. The following specific measures are suggested.

(1) Greater financial incentives to expenditure on energy-saving schemes.
(2) Use of existing authority to strengthen the requirements for thermal insulation in buildings.
(3) The enactment of an Energy Bill, with provisions inter alia for:
   (a) the introduction of an energy licencing scheme to directly control energy consumption in commercial and industrial buildings;
   (b) amending relevant Acts so that all nationalised energy industries have a statutory duty to take account of their national energy considerations as well as their particular industry;
   (c) empowering the Minister to amend other legislation which is found to have the practical effect of hindering energy conservation;
   (d) giving general powers to enforce greater co-operation between energy industries and to restrict the extent and form of advertisments for energy;
   (e) to set up an energy advisory service for householders.

REFERENCES

1. Factories Act 1961, s 3–5.
2. Offices, Shops and Railway Premises Act 1963, s 6–8.
3. Standards for Schools Premises Regulations 1959.
4. The Fuel and Electricity (Heating) (Control) Order 1974.
5. The Thermal Insulation (Industrial Buildings) Regulations 1958.
6. The Building (Second Amendment) Regulations 1974.
7. The Building Regulations 1972.
8. 'Domestic Engineering Services'—Report published by IHVE 1974 (the Billington Report).
9. Billington, N. H. (1974). Thermal insulation of buildings. *Building Services Engineer.*
10. Hasler, A. E. (1974). Why waste all that heat? *The Consulting Engineer.*
11. Betts, P. E. and Boley, T. A. (1969). Electricity alone can supply all the consumer's energy requirements *Electrical Review.*
12. West, G. S. (1975). Electricity supply industry's role in district heating *The Heating and Ventilating Engineer.*
13. Address by the Secretary of State for Energy to the Association of British Science Writers, 26th June 1974.
14. Hustings, J. (1975). Assistant Planning Officer (Building Control), Breckland District Council. Letter to the *Architects J.*, 26th March.
15. Circular letter from the Department of the Environment to Local Authorities 29th April 1975.
16. Technical Information Note on Revised Part F—*Thermal Insulation,* The Building (Second Amendment) Regulations 1974, Department of the Environment Welsh Office.
17. Owens, P. G. T. (1975). Energy budgeting *Building Services Engineer,* March.
18. Johnson, W. H. (1975). A need for compulsion *Building Services Engineer,* May.
19. *Energy Conservation in the United Kingdom* HMSO, 1974.
20. Reservoirs (Safety Provisions) Act 1930.
21. Nicholson, F. J. and Fitzgerald, D. (1975). Volumetric heat losses *Building Services Engineer,* March.

Discussion

F. A. Pullinger (*Haden Carrier Ltd*): This conference is sponsored by the Department of Energy amongst others. Mr Jamieson asked some very pertinent questions. Is the Department able to give us some indication of their thinking on these questions? Of course they are very difficult to answer as the paper by Mr. Jonas, which we will hear later in this conference shows. Nevertheless we need to have some indication of the life of fossil fuels and the time taken to provide new generations of nuclear stations. We have been told that the industry must advise our customers of the best way of solving their problems. Should we now decide to use fossil fuels for say the next 15 years.

Now on a second point, we can easily, as designers or consultants, give advice to our customers either on the straight forward 'pay-back' basis or on the more sophisticated basis of 'discounted cash flow', about relative energy costs. What of course we do not know is what policy changes might take place. The Government cannot predict them entirely but there are certain elements that they can give a line on. Will the relative costs of fossil fuels be maintained at some fixed level.?

Mr. Swain has told us about his ideas on legislation which are very interesting and very important. It is easy to ask for legislation but the Government who has to practice these things and monitor them will have a better idea than us how far it is practical. If we could get some answers from the Department of Energy as co-sponsors of this conference, I think we should all be most grateful.

Dr. P. V. L. Barrett (*ICI Insulation Service Limited*): This conference is concentrating rightly on the design of buildings but there is one point I would like to make referring to Table 5 in Mr Swain's paper. I think it is fundamental in terms of Government legislation. He has given a column in the table stating the Relative Gross/Net Energy Ratio. I would submit that he has given the Primary/Gross Energy Ratio, which is the energy delivered to the house.

Now what is not considered here, and which is very important to the domestic sector, where central heating appliances are a lot less efficient than in industry, is their annual operating efficiency. Taking figures from a paper of my own we now have:

	Relative primary/ gross energy ratio	*Annual operating efficiency*	*Relative primary/ nett energy ratio*
Coal	1·00	50%	1·00
Natural Gas	1·04	63%	0·83
Oil	1·07	65%	0·83
Manufactured fuels	1·37	63%	1·09
Electricity	3·65	90%	2·03
Self-generated (solar, etc.)	0		0

I feel in some senses that the vendetta against electricity has gone a little too far and the above table gives a better perspective of the relative performance of fuels.

However, what the annual operating efficiency figures show much more clearly is the huge energy waste due to inefficient appliances. Appliances can be designed to give a higher annual operating efficiency. The annual operating efficiency must not be confused with the bench efficiency of say, an oil fired appliance of about 80% which can only be achieved by continuous steady firing to achieve thermal equilibrium, taking many hours. This condition is never achieved in practice. I therefore feel that this is an area where research effort must be put in.*

* Barrett, P. V. L. 'Better comfort at a lower cost' *Build International*, **7**(529). 1974.

C. P. Swain: Thank you, Dr. Barrett, for pointing out that Table 5 is of primary and secondary energy ratios. My figures and their description were taken from a Government publication—'Energy Conservation in the United Kingdom'—and there is a clear need to standardise nomenclature in this direction.

In my view the object of this type of energy limitation should be solely to limit the amount of primary energy a consumer may use and give him complete freedom of choice as to how he uses it. It must be the customer's decision as to whether to take this primary energy in the form of a basic fuel or as a more sophisticated form of energy. Provided the consumer keeps within his primary energy allocation the question of differing appliance efficiencies is not material.

I. Munro (*Eurisol, UK and Association of British Manufacturers of Mineral Insulating Fibres*)*:* Whilst we would obviously pay high tribute to the Department of Energy for their 'Save It' campaign, Mr. Jamieson and Mr. Swain have both referred to exhortation, education and example, as being admirable, although not enough; effective enforcement in new dwellings and encouragement by incentive in existing buildings are vital.

Mr. Jamieson mentioned two countries where incentives were now being paid, Sweden and Holland. In addition in Europe incentives are now being paid in Denmark, Belgium, France and Germany. Germany had previously only held a competition for the best ways of saving energy but they have recently announced a major once for all improvement grant scheme which makes particular play of energy conservation measures by subsidy in a total for the whole improvement scheme of 700 million D. Marks. France has a fiscal scheme which can allow up to almost £800 per dwelling in reduction of income tax and which makes great play of improvements in energy conservation measures. Although we have our financial difficulties in this country we are only one among several nations in Western Europe who have the same problems and when the Government has spent £5 million within fifteen months on exhortation we could perhaps have some incentive however small particularly to householders to insulate.

J. M. Cooling (*Balfour Kilpatrick Ltd. President—Institution of Heating and Ventilating Engineers*)*:* Many of the contributions have really been questions directed to the Government or the Department of Energy. Perhaps someone from the Department of Energy would like to comment on the questions which have been raised.

J. C. Denbigh (*Department of Energy*)*:* It is invidious for a Civil Servant to get involved in the policy of his Department so I can only try to answer the questions which have been put from a technical point of view and not deal with the political aspects.

Mr Swain very delicately said 'let the Government put their money where their mouth is'; this was a charming way of asking the taxpayer. The Government does not have money, it is the money of the people who are at this Conference, it is the money of all taxpayers. What Mr. Swain was saying is 'let all the taxpayers put their money where the Government's mouth is'. I suggest that there are no free handouts and everything we do as a Department must be for the good of the country and the taxpayer as a whole, and if we're looking for incentives in the consumer fields, the present policy is—if it benefits the consumer financially, then it is in the consumer's own interest to do it. We will assist him in making the decisions. We will advise him as far as we can what to do. When it comes down to the finance we have an $11\frac{1}{2}\%$ interest loan scheme for energy conservation measures in industrial buildings but we are told this is not enough. What we are being asked now, presumably, is to ask the taxpayer for more income tax to pay for somebody's energy conservation. Our policy is not to do that. I am not saying this policy will not alter.

Combined heat and power is a very old question. The first combined heat and power station was, I believe, built about 1904 and linked with a district heating scheme in Manchester. These schemes have had varied lives and with the increase of interest in energy conservation, the Department of Energy is now looking at this subject. Mr. Jamieson mentioned the committee looking at restraints on electrical generation. This is principally concerned with the relative exchange rates for private generators. We also have a committee under Dr. Walter Marshall, looking at the whole question of combined heat and power in the UK which includes long range heat transmission from large nuclear stations. A report will be published in 1976, it would obviously be premature to discuss the findings now.

There has been a lot of discussion about fuel pricing, what one should charge for oil or gas which are indiginous fuels, where in effect we are not subject to market prices from outside. There has been consideration of pricing on a cost per therm basis. This has a number of problems as the capital costs and operating costs of different types of plant obviously contribute to the actual cost in use.

C. P. Swain: I appreciate that it is our money I was talking about. My point is that I should like to see more of my money being applied to energy incentives at the expense of other ways in which it is being spent at present.

As a Civil Servant, Mr. Denbigh must remain outside party politics—and we all respect that. I believe energy itself should be outside party politics. But Mr. Denbigh has given a politician's answer in talking about the need to raise income tax to pay for the energy conservation incentives I have advocated. There is no need to raise any taxes if expenditure can be saved in other directions. It is a question of priorities, and at present I believe we have our priorities wrong so far as the expenditure of my money is concerned.

N. S. Billington (Building Services Research and Information Association): Dr. Barrett referred to appliance efficiencies and the need for work to be done upon this. Now quite recently the EEC have invited tenders for research projects on energy conservation and energy-saving devices to the extent of 11 million units of account. This is something like £6 million. One of the proposals we are submitting relates exactly to the work Dr. Barrett has mentioned. I would like to ask whether the government is prepared to assist

UK research organisations, consultants and others to attract some of this money, or will it, as I fear, reduce its own contribution to these same organisations if they are successful. On a second point, Mr. Swain has given some fuel factors. I think one ought to be careful in discussing the electrical ones. For certain uses electricity is the only feasible source of energy for lighting, for household appliances, for television etc. For these purposes it would be unfair and unwise to put a very high factor upon electricity, and, finally, may I return to something that Mr. Denbigh has just said, and here I go along with Mr. Orchard. Insulation of a dwelling can yield a capital saving for energy production (whether in North Sea oil or gas or electricity generation) which is much greater than the cost of the insulation itself. The tax payer benefits by 100% grant for insulation of dwellings.

C. P. Swain: As Mr. Billington points out, there is much discussion at present on the form that primary/secondary energy weighting factors should take.

Mr. Billington rightly says that for certain uses electricity is the only feasible form of energy, and that account should be taken of off-peak electricity costing less in terms of primary fuel than on-peak electricity. Consideration should perhaps also be given to the fact that electricity may also be generated from fuels that are unusable for other purposes. But my belief is that energy weighting factors must be simple to apply. So I prefer to take account of every user requiring some electricity for lighting and power in the allocation of primary energy rather than attempting to distinguish between electricity used for differing purposes.

J. J. Ballance (Allied Irish Banks Ltd.): As an Architect myself, I would like to support energy conservation legislation because I believe it would enforce a certain conformity on the architectural profession. I find a great difficulty over recent years has been that the engineers have been giving very positive advice on energy conservation but quite frankly it has been filtered by architects and does not get to the clients in the form that it should—I think that at least legislation enforces the engineers expertise on to clients past the filtering medium of the type of architect who designs an energy wasteful building and demands that the services consultant then produces a habitable internal environment.

J. C. Denbigh: I speak on behalf of the Department of the Environment. We have had criticism from Mr. Swain on the efforts of the Building Regulations Division, now the call from Mr. Ballance. It must be realised that legislation can usually do no more than set down a minimum which approaches the economic minimum. In practice we should try to do better, try to build above mandatory standards—when Mr. Orchard made the point about the £300 housing cost yardstick for district heating I felt I would rather see that £300 or at least part of it, spent on insulation. However at present there is no sanction for higher standards of thermal insulation, and Local Authority dwellings will be built to the Building Regulations or possibly a little bit better but without the ability to divert money from what might be the district heating side of the scheme to have higher standards of thermal insulation. In other words I wholly support the point made by Mr. Billington.

G. D. Nash (*Building Research Station*): May I refer to points made in relation to insulation of housing. Doubtless many people here realise that the regulations involving standard of thermal insulation have not been made primarily for energy conservation purposes but for health requirements including the prevention of condensation. The 1974 Act provided authority for making regulations specifically for energy conservation.

I was interested in Mr. Swain's comments about the effect of size and shape on thermal conduction loss. Satisfying the regulations does not necessarily provide the optimum design solution—I think that is the job of the architect and the consultant; the regulations as has been indicated, tend to give a minimum requirement. If regulations were to cover all design situations in practice, they may become very complex—different energy standards may be desirable for different types of house design. A minimum energy requirement for a dwelling would favour say blocks of flats as was illustrated in Mr. Swain's paper. Various design aspects have to be considered by those making regulations.

I would like to comment on the progress of thermal insulation in housing. In a paper presented to the Institute of Fuel Conference in 1956* I compared the 'house U value'—the overall rate of conduction loss from the external enclosure of a house—for a pre-war house, a post-war house, a future-house[1] and a future-house[2]. The 'house U value' for 'future-house[1]' is substantially achieved with the thermal insulation standards in the current regulations (this does not involve the question of size and shape). Considerable progress since pre-war has been made but there is still scope for improvement.

J. W. Weller (*Adams Green & Partners—Consulting Engineers*): The majority of Building Designers and Consultants are aware that the current building regulations are based upon the Public Health Act and not upon the Principles of Energy Conservation. Nevertheless, there still appear to be inconsistencies in the legislation for the thermal performance of buildings.

Certain Physical requirements are legislated for dwellings to safeguard the health of the occupants. Yet when these people leave their dwellings for offices or schools they move into buildings which are in the majority immune from any legislation on the thermal performance of buildings. Hospitals, where health is surely of prime importance, are a notable example.

Previously these inconsistancies have been of little concern as buildings are heated to ever increasing temperatures and excessively ventilated to alleviate detrimental effects to health. This is an energy wasteful solution.

I would suggest that the legislation governing the thermal performance of buildings be re-examined to remove these inconsistencies and, at the same time, introduce legislation to improve energy conservation by our new buildings.

* Broughton, H. F. and Nash, G. D. (1956). 'Structural Insulation and Heating Systems—Values and Capital Costs'. Conference on Special Study of Domestic Heating in the United Kingdom. Institute of Fuel. pp. 229–237.

C. P. Swain: To answer Mr. Nash—some of the points I have criticised in Section F of the Building Regulations have no doubt arisen because it was not prepared (at least openly) as an energy saving measure. But my criticisms were aimed at demonstrating the dangers and pitfalls of prescriptive legislation of this type.

Of course we should be designing to better than the minimum standards prescribed by legislation, and I am sure that professional architects and engineers will endeavour to do this. But with the pressure of cost limits etc. it is in my view probable that in the majority of cases the minimum standards will tend to become the norm, particularly so with dwellings that are not architect-designed. I am very much in agreement with the comments made by Mr. Weller.

Mr. T. Hedgeland (Electrical Contractors' Association): May I draw attention to what seems to me to be an anomaly between two features which have arisen during the discussion. On the one hand we are talking about the energy situation at the turn of the century, or 30 years ahead, whilst on the other hand we are talking about the well-being of our grand-children, in which sense we should be talking about a time of, say, 60 years hence. In particular, this would ally with the life of buildings as now designed, generally being of the order of 50–60 years. Hence we should be looking further ahead and perhaps the Building Regulations should, at this time, have much more stringent requirements with regard to thermal insulation and other energy-saving aspects of services in buildings.

The principle of the autonomous house is significant from the point of view of energy, as exemplified by the current investment, in Europe, by the Philips organisation in their 'autonomous' house at Aachen in Germany.

Again, from the point of view of 'low-energy' housing, when considering the two factors of energy use involved, high-energy materials use in construction are a small part of the content compared with the on-going use of energy during the life of the building. It would therefore make sense to invest more money in the near future in 'high-energy', energy saving materials, to save a far greater amount of energy during the on-going life of the building. In this sense, the current philosophy of 'live now—pay later' will have to be reversed to one of 'pay now—live later'.

G. Nelson (Oxford Polytechnic): Mr. Swain's and Mr. Jamieson's papers both support my view that proposed projects for the conservation of energy are hamstrung by accountancy, for the cost of them must not have an adverse effect upon profitability.

I suggest that any feasibility study must not be based on present day prices but upon projected costs in the year 2000. This conference is not about cash flow based upon present day conditions it is about the possibility of survival into the next century.

A. Gordon (Alex Gordon & Partners): I would like to say something—in response to Mr. Ballance—on the comparative performance and responsibility towards energy conservation among engineers and architects. I think it would be very difficult to assess this numerically. In both of the specialisms there have been people who have been thinking about the subject very hard, but they are small in number. I like to feel as an architect that we were thinking a little earlier than some of the engineers but I may be entirely wrong. You will be hearing other architects speaking about their efforts during the conference. One thing I think, however, is clear; that we have come to the end of a period when the bad architect is bailed out by the services engineer, where bad buildings are made habitable by terribly expensive systems which are very expensive in energy and monetary terms. I think this is a very important point, and is linked with the question of fees. When buildings are themselves considered as environmental systems the cost of the services installations may well be much lower than they are at the moment, and this is where the professionalism of the engineer will come in. He will have to put in more effort and at the same time reduce his fee. I have said in other places and I will repeat here that I think the services engineer's fees are far too low. The whole basis of professionalism is that the effort put in is that which is necessary best to serve the interests of the client, not your own pocket. It is, however, unreasonable to expect such an approach if what you are being paid is unnecessarily low. It is this situation which discourages the search for economy and encourages leaving design aspects to the contractor. Proper fees and a professional approach will lead to better overall designs and economy in energy use.

3

Spending Money to Save Energy

BERNARD LUBERT

Synopsis

The paper is based on the experiences and plans of Marks and Spencer Ltd. It provides information on the size and scope of the organisation and details of the Energy Audit at the time of the 1973/4 Crisis. The author describes the immediate steps that were taken following the crisis, the new developments in lighting that have been employed and the plans for Capital expenditure on lighting, refrigeration, heat recovery and building insulation.

Marks & Spencer and Energy Conservation

The information contained in this paper is drawn from the experiences and plans of Marks & Spencer and it would, I think, be appropriate if I gave a little background information on this organisation.

Marks & Spencer has a total of 252 Stores throughout the UK with a sales area of $5{\cdot}3 \times 10^5$ m^2 (an average of about 2137 m^2 per unit) selling mainly textiles and foods. It has recently embarked on some development in France, Belgium and Canada in addition to an annual increase in this country of between 5% and 10%.

The Company's net turnover in 1974/75 was £721 million after giving refunds of £45 million as part of its well-known policy, and deducting £39 million in VAT. The gross operating profit was £81 million and the net, after taxation, was £39½ million. The total cost of energy during the year was £2 928 000.

You may well think that, comparatively, this is a small figure and wonder why so much time and effort have been devoted to Energy Conservation measures.

The reasons are twofold. Firstly, it became apparent that without the necessary steps being taken, the cost of energy would have become a far more significant figure in the running cost of a business operating under strict statutory limitations on its gross profit margins. Secondly, it is deemed a duty to the country in which it has grown up and prospered, that every effort be made to save energy and to assist in a National Campaign on Energy Conservation which is vital to the economic well being of the country and therefore of us all.

THE FIRST STEPS

The events at the end of 1973 and early 1974 convinced us that never again would energy be cheap and abundant.

The period of Industrial Action had provided us with experiences upon which we were able to draw.

We learned that we could operate very satisfactorily and, possibly, a little more comfortably, at levels of illumination well below those we had thought necessary before.

We commenced a study of all uses of energy, and related, more specifically, throughout our business, overheating and overcooling to waste, and instituted Energy Management.

Our first steps were to reduce our illumination levels by over one third—from 1000 lx. to 600 lx. This was facilitated because the majority of Stores had arrangements that permitted us to switch off a number of the tubes in each of our suspended fluorescent fittings.

We had long before decided against the use of switched start tubes and, with the control gear we employ, there would have been no saving by removing tubes. We therefore embarked on our earliest expenditure by dealing with those larger Stores without part switching facility, by installing switches within the light fittings. The expenditure was about £1300 per Store which gives now a saving in electricity of £2700 per annum.

All the 252 Stores were visited within a period of less than a month by members of the Company's Building Division to provide on-the-spot guidance to Store management on the operation of Energy Management in lighting, refrigeration, power and heating—an expensive but worthwhile exercise.

The very successful process of creating and then maintaining a new attitude of mind, indeed an enthusiasm, by all Staff towards Energy Conservation was commenced.

The Programme

In April 1974, following the submission to the Board of Directors of a Report on Energy, the programme for the measures to be taken was decided upon.

The report provided the following information:

Energy Costs for the year 1973/74 had been as follows:

Electricity £2 033 000 and Gas £381 000, Total £2 414 000 with adjustments for generation of emergency power.

Electricity costs were incurred in the following approximate proportions:

Sales Floor Lighting	£700 000	35 %
Foods Refrigeration	£600 000	30 %
Air Conditioning, Ventilation & Electrical Components of Heating	£450 000	$22\frac{1}{2}$ %
Other lighting and power	£250 000	$12\frac{1}{2}$ %

Gas provided the Heating, Hot Water Supplies and Catering.

It was quite clear from the above figures that the most immediate and significant reductions could be effected in the field of lighting. The first measure was to reduce all levels of illumination from 1000 lx. to 600 lx. but it was apparent too, that more efficient light sources could be employed.

SALES AREA LIGHTING

Almost all Stores had fluorescent lighting contained within suspended luminaires fitted with 'egg crate' type louvres to reduce glare. The utilisation factor of the luminaire is about 0·55. Within the luminaires Natural fluorescent tubes were installed being either 1·52 m. 80 W., or 2·44 m. 125 W.

It has always been considered essential for the Sales Floor lighting to have the best possible colour rendering properties, within reasonable commercial limits, and the Natural fluorescent tube had been developed about 20 years ago to meet this requirement. It was, however, an unfortunate fact that the efficiency of fluorescent lighting, in lumens per watt, dropped inversely with the colour rendering properties.

Fortunately, we learned from Philips Lighting Ltd. that their phosphor chemists had been re-examining the spectral distribution requirements of fluorescent lighting to provide good colour rendering and had reached the conclusion that the highest efficiency could be achieved by concentrating the energy in the wavelengths of the three primary colours, red, green and blue. By the development of phosphors containing the rare earths Eu^3 (red) trivalent europium, Tb^3 (green) trivalent terbium and Eu^2 (blue) divalent europium they were able to produce fluorescent tubes, which they call TL. 84, with a very high colour rendering index of 85 (against 80 for the Natural tube) and a light output, including gear losses, of 56 lm. W^{-1}. compared with 40 lm. W^{-1}., 40 % more. The colour temperature is similar to that of the Natural tube at 4000 K.

Our calculations showed that despite the very high cost of the TL. 84 (well over 5 times that of the Natural tube), by reducing by 30 % the number of tubes installed, there would be an overall saving in cost of £700 per annum per original 1000 tubes.

This calculation was based on an average cost of electricity of 1·1p/unit. The saving will be considerably greater at the present cost of about 1·9p/unit.

It was agreed to try out the TL. 84 tube in the upper Sales area at Lincoln Store which was being developed at the time. Both 1·22 m. and 1·52 m. tubes were used and the readings taken bore out the earlier calculations. We have worked very closely with Philips on this development and would like to take this opportunity to thank them for enabling us to be the first organisation in the world to carry out large commercial installations with this exciting new development.

The decision was taken to re-tube the business over a three year period with TL. 84 tubes, at the same time disposing them equally within the luminaires, disconnecting redundant gears and cleaning the luminaires. We have the new tubes in four new and extended Stores and in 54 other Stores, including our most important one at Marble Arch. A hundred Stores will be so equipped by April 1976.

Parallel with this we had been interested in a compact source form of lighting which we had tried at our Stores in Kentish Town and Norwich and had planned for our new Store at Ayr in Scotland. The efficiency of mercury halide lighting was comparable with TL. 84 installations, and, while I must apologise for departing slightly from my brief on spending money, installation costs of these simple luminaires were considerably lower than our fluorescent fittings. Our textile technologists were not entirely happy with the colour rendering properties but Thorn Lighting who, again, have worked very closely with us, have now produced the required result and it is our policy to install mercury halide lighting in all the new Stores we are building. We should not, however, mix this form of lighting with fluorescent and therefore any extensions to existing premises will be matched with fluorescent luminaires containing Philips TL. 84 tubes.

We have always found it a necessary part of the Marks & Spencer sales area image to illuminate the perimeter of the area with fluorescent lighting behind a pelmet. This was done using fittings carrying, usually 1·52 m. and 0·61 m. tubes controlled by instant start gear. The total length of tubes employed was 70 104 m. of which 34 138 m. was in twin form.

It was decided to change all the twin tube fittings to single tube which involved considerable alterations to gear, including the replacement of 6000 chokes for 0·61 m. tubes costing £24 000, to eliminate 17 069 m. of fluorescent tube at a total cost of £90 000. The estimated annual saving at the time was £70 000 on electricity, apart from periodic replacement costs of tubes, but this will now be well over £80 000. The programme is now complete.

FOODS REFRIGERATION

Our foods business in 1974/75 totalled £211 million of which £92 million was on chilled or frozen foods. At the time of the report about 30 % of our electricity bill was incurred on foods refrigeration, making it the second largest item. The considerable savings on Sales Floor lighting costs which were immediately effected and the ongoing programme of improvement, together with the continued expansion of our refrigerated food business have now placed this item of consumption well into the lead at about 34 % to 35 % of our total electrical costs.

The first steps taken were to institute immediate economies in operation by condensing displays in the early part of the week and switching off as much equipment as we could without affecting sales.

The next step was to examine the equipment we were using for operating efficiency. The two most popular items were (a) the chilled food multi-tiered 'gondola' specially developed for Marks & Spencer to provide the most simple accessibility to the public and maintain our stringent temperature requirements and (b) the multi-tiered frozen food side counter. Both of these units are operated from compressors and condensers remotely installed, often on the roof. Both are very costly to run.

As far as the former is concerned the design has been modified, after experimental alterations carried out at a cost of £300 (4·88 m. gondola) to reduce the possibility of spillage of cold air, albeit customer access is slightly restricted. Some units are being modified and all new units, of course, contain the modification. We have also evolved a new form of island unit, which we call 'Slimline' which is made up of 'mobile' units (incorporating their own compressor/condenser units) and which complements the 'gondola'. It is our belief that a mixture of 'mobile' and 'piped' units permits us to achieve a balance between heat loss and heat gain and work towards the most efficient use of the energy consumed.

While we have retained the existing multi-tiered frozen food units we have discontinued their use in further installations. Our equipment suppliers, George Barker of Leeds, have developed, with us, new tub units, which require far less energy to run, since they retain their cold air, lend themselves to the simple application of night covers and, despite their apparent lesser accessibility, appear to perform equally well on sales. The provision of night covers, at a cost of about £15 000 will result in an annual saving of about £20 000 on electricity. Considerable research work is still progressing on developing suitable night covers for the tiered units but we have not yet been fully successful, to a large extent because we operate our chilled units at lower temperatures than the general trade and are therefore closer to freezing conditions; and we have problems with short cycling of the compressors on the frozen units. However, progress is being made.

AIR CONDITIONING AND VENTILATION

The reduction of our Sales Floor illumination levels has provided us with a bonus on our running costs in this field. Lighting constituted over 30 % of our total cooling load and our automatic capacity control on the refrigeration equipment will already be providing us with savings.

However, the greatest usage of electricity is incurred in the fans employed to move the air. We are embarking on a programme of reducing the air volumes and therefore the horse powers required by simple restriction of the inlet eyes of our centrifugal fans. Design work and costings are being completed at the present time.

The experiment carried out at our High Wycombe Store showed that an annual saving of £450 would result in a Store of 2600 m^2.

OTHER LIGHTING AND POWER

There had earlier been an attitude that incandescent lighting was more comfortable in offices and staff quarter areas. The energy crisis brought home to us that fluorescent lighting is over 3 times as efficient. Old prejudices were quickly forgotten and designs for all new installations were prepared on the basis of fluorescent lighting utilising the efficient white and warm white tubes where colour rendering was not important. We are embarking on a gradual and long term programme of change but it is interesting to note that many luminaires designed some time ago for tungsten lamps can be adapted to take circular fluorescent tubes to give a far better appearance and much lower running, and bulb replacement costs.

We are, in our Stores, engaged in a programme of installing Power Factor Correction equipment where the tariffs are appropriate and have shown that the installation pays for itself in two years.

We are also installing Maximum Demand warning systems in two Stores in order to evaluate the systems before carrying out a wider installation programme.

GAS

Costs for heating, hot water supplies and catering had risen from £237 000 in 1964 with a sales area of $2{\cdot}97 \times 10^5\,m^2$. to £469 000 in 1974/75 with a sales area of $5{\cdot}11 \times 10^5\,m^2$.

This small comparative increase, bearing in mind recent price increases over ten years was probably due, to an extent, to four factors:

(a) A very mild winter last year.
(b) The setting down of our thermostatic controls from 20 °C to 18·3 °C, together with the efficiency of our thermostatic controls in the upper two thirds of our Stores containing the larger units.
(c) The reasonably good standard of insulation of these more modern buildings.
(d) Our rudimentary Heat Recovery arrangements in existing systems.

Nevertheless, we anticipate considerably higher costs this year in budgeting for a normal winter (we have already had a very cold start to our financial year), for lower 'spin-off' heat benefits from lighting, and the extremely high increase applied for by the British Gas Corporation.

We are endeavouring to introduce measures at least to contain the rise in our developing square footage and to apply additional thermostatic control in those Stores where there is room for improvement.

We shall re-use the heat released by the foods refrigeration process with:

(a) More 'mobile' equipment.
(b) Installing duplicate condensers within our stockrooms in place of conventional unit heaters,
    or,
(c) Employing damper control systems on the condensers to divert the warm air to the Stockroom in the winter or discharge it in summer.

The two latter arrangements will add about £2000/£3000 each to the cost of our installations which will be partly offset by the use of some integral 'mobile' equipment.

We have included in our Budget a sum for fitting thermostatic radiator valves to all radiators not already so fitted. We have also allowed for modifying the control arrangements to heater batteries on air conditioning and plenum plants to prevent any loss of heat on switching off, and for more directly regulating the control of pumps to cooling and heating requirements.

We are designing into our new installations a more direct recovery or rejection of heat by ducting direct from our light fittings. This is costing over £10 000 per installation in additional ducting.

We shall be carrying out an installation this year at Weymouth of solar panels for the Store's domestic hot water requirements.

We are now incorporating into new Stores the use of cavity wall insulation using mainly the Pilkington product 'Dritherm' at a cost of 99p/m^2 supplied and fixed and are standardising on double glazed windows in Stockrooms, Offices and Staff Quarters at additional costs of £2000 to £5000 per Store. In these respects we are working very closely with Pilkingtons.

The 'pay-back' period of these items may be over 10 years at present prices but this may well shorten considerably with projected price increases. For this reason alone we are sure that this is the correct policy and, again, the case is strengthened by the moral requirement.

Summary

Apart from money we are spending on our Development projects we have budgeted to spend this year nearly £$\frac{1}{4}$ million, and in each of the next two years, on Energy Conservation measures, including Research and Development.

This will be fundamentally on lighting and lighting controls, on foods refrigeration improvements, on modifications to our air and water moving arrangements in air conditioning and heating plants, on certain heat recovery arrangements and on improvements to structural insulation.

The easy part on saving energy by switching off is now over. Further improvements can only be achieved by careful examination of the design, function and efficiency of all energy using equipment together with the buildings that it serves. This must be followed by a programme of progressive investment to improve performance, with the application of commercial good sense in consideration of 'pay-back' period, tempered, perhaps in some cases, with the social obligation to assist in an overall effort so badly required by the Nation.

Discussion

P. R. Ralph (*Unilever Research*): I would like to hear Mr. Lubert's comments on the part preventative maintenance played in his plans to conserve energy. I would also like to hear how he views the economics of annual performance checks of re-commissioning of major plant (say air conditioning plant) with respect to energy savings.

B. Lubert: We have not actually evaluated the effects of planned maintenance on energy conservation measures but we do have a planned maintenance programme which is pretty detailed and closely specified. The question of the performance checks for instance on boiler plant or air conditioning is something one might well look into.

J. B. Collins (*Building Research Establishment*): I understood that in supermarkets 1000 lx. was needed to encourage customers to come in and to buy—has Mr. Lubert any evidence that going down to 600 lx. has resulted in any reduction in sales?

B. Lubert: Slightly to the contrary, we found that when we reduced the lighting levels we had some very complimentary remarks from the customers and certainly a considerable proportion of the Staff, on whom we rely to a great extent for our level of sales, were also very happy at the reduced level of lighting. This is something the period of difficulty during the industrial action in 1974 taught us following a time when we had thought that the higher lighting levels were going to make an impact on sales.

W. R. Cox (*Heating and Ventilating Contractors Association*): It has often occurred to me that department stores are heated for the benefit of the assistants rather than for the comfort of the customers and are generally badly overheated in relation to customers who are dressed in outdoor clothes, while in contrast the assistants are often scantily clad! I wonder if Mr. Lubert could comment on this and on the practicability of encouraging assistants to wear more clothes so that temperature can be lowered.

B. Lubert: As I said before we rely on our sales assistants to provide us with a high level of sales. It always has been a difficult problem to find a temperature that is satisfactory for both staff and customers. Assistants are there throughout the day, the customer for an average of 20 minutes or so. But I did say that we had lowered our temperatures to 18·3 °C. It was difficult in the days of more intense lighting to maintain those temperatures during Autumn and Spring because the lighting tended to raise the temperatures. In addition, in a Store of our type, or a supermarket, you do get local areas of cold where you have a large bank of doors into the Store which requires local heating. That heat tends to get wafted at times into the main body of the Store where incidentally we do not provide heating. It is a difficult problem; we aim at lower temperatures, I do not think we always get them. In our publications to our staff, we have asked them to put on warmer clothing. Generally speaking, there have been no complaints at the lower temperatures at which we have been working.

D. N. Soane (*wrote*): Mr. Lubert gave details of the savings made by Marks and Spencer Ltd, in conserving energy. The overall result, in monetary terms, clearly showed it is desirable in many instances, to spend money, to save money. Whilst appreciating the necessity of economic design and use of cheap materials, much can also be done by installing many of the energy saving devices discussed during the Conference, saving even more energy and money in the longer term. This requires a more realistic approach to cost limits to ensure that this can be done during initial design, and that wasteful 'economies' imposed by very close, and in some instances outdated cost limits, no longer occur.

4

Energy Conservation Measures in the National Bureau of Standards Laboratory Complex

P. R. ACHENBACH and D. A. DIDION

Synopsis

Energy conservation measures at the National Bureau of Standards site are being carried out under two separate but inter-related programmes; a low-investment, immediate-impact programme and a long-range, major retrofit programme. The low-investment programme has already been implemented and includes lighting reductions, thermostat resettings and night time shutdowns of heating, ventilating, and air conditioning systems. Data on energy usage before and after this programme was instituted are presented. The methodology and analysis of the long-range programme are also presented. Various building energy conservation options were considered and evaluated by means of a mathematical model. Quantitative estimates of savings for each are presented. The decision to invest in automatic controls for the air conditioning systems of the laboratory complex and steam/chilled-water power plant is discussed.

Introduction

The following presentation is concerned with the energy conservation programme for the operation of the plant and buildings at the Gaithersburg site of the National Bureau of Standards. Because of the unique nature of this conference, some of the more generic aspects of retrofitting buildings for energy conservation will be mentioned with the experiences of this study acting as examples which emphasise the problems.

The general aspects of a building retrofit procedure that will be exemplified are:

(1) The need for a histogram of base-line energy data which enters

the site, against which results of conservation innovations may be measured.

(2) The need for analysis of energy data on a subsystem basis under conditions of actual operation, since design data are often inadequate and misleading.

(3) A sequential approach of looking at building operations, then controls, then other hardware alterations, based on the logic that lesser initial investment ideas are more readily accepted by management.

(4) The need for a careful engineering application of the 'popular' energy conservation ideas to the specific building and location, to confirm cost benefits, practical ramifications, and energy savings.

(5) The need for a sophisticated engineering analysis of the complete building system for synergistic effects where major renovation is intended.

Description of Site and Systems

The relatively new NBS Gaithersburg laboratory complex is a flexible facility comprising 23 buildings with more than $2{\cdot}1 \times 10^5\,m^2$ ($2{\cdot}25$ million ft^2) of floor space (see Fig. 1). These buildings are spread over a campus of some $2{\cdot}33 \times 10^6\,m^2$ (576 acres) and use an average of 115 million kWh of electricity and $8{\cdot}23 \times 10^{14}$ J (780 billion Btu) of heating fuel each year (5929 MJ/m^2 (153 kWh/ft^2 yr) of energy usage). Approximately 2000 individual laboratories house highly precise and often delicate scientific apparatus which often requires fume exhaust systems and clean chambers. In addition, there are specialised laboratory buildings which house major facilities such as a nuclear reactor, wind tunnels, linear accelerator, a $5{\cdot}4 \times 10^6$ kg (12 million lb) mechanical dead weight tester, and a paper mill. As a general rule all the occupied space has been maintained at 23·4 °C (75 °F) and $< 50\,\%$ relative humidity and has required, on the average, 472 m^3/s (1 million cfm) of outside air.

Each of the buildings receives energy in two forms, thermal and electrical. The thermal energy is supplied in two modes, high-temperature saturated steam at 11·4 bar (150 psig), 177 °C (350 °F) and low-temperature chilled water at 5·6 °C (42 °F). Generation of the steam and chilled water takes place in four steam generation units

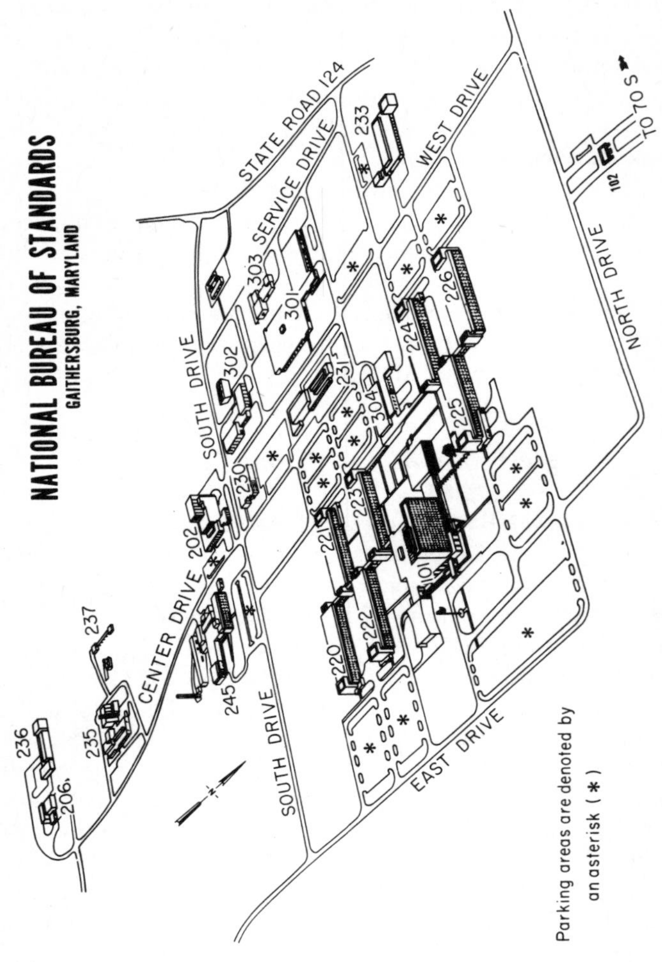

Fig. 1. National Bureau of Standards.

of 5·5 kg/s (44 000 lb/h) each and four centrifugal compressor refrigeration units, of 2237 kW (3000 hp) each, respectively, located in the plant building on the NBS grounds (see Fig. 2). The electricity is supplied through the local utility power grid network (which services most of the population of Northeast United States) by way of a substration on the NBS grounds.

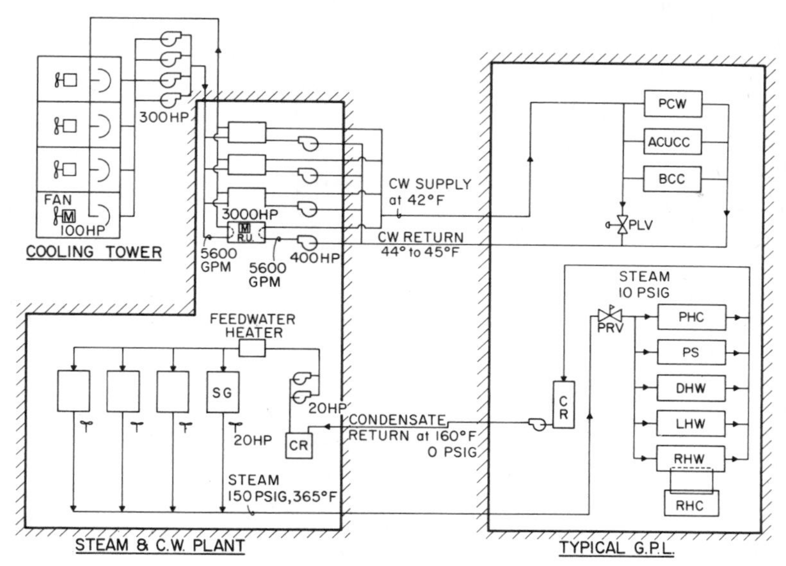

Fig. 2. Steam and Chilled Water Distribution System.

Natural gas is the primary fuel for steam generation; however, the Washington Gas Company may, upon a 2-hour notice, interrupt the gas supply if there is an exceptional surge in demand. At such time, the standby fuel, No. 2 oil, is used. The underground fuel oil storage capacity is 568 m^3 (150 000 gal), which is approximately 7 to 10 days supply. During the winter of 1974–75, 40 % of the heating energy was supplied by oil. Although steam is used as process steam in the laboratories and shops, and for domestic and laboratory hot water, the vast majority of this energy (> 80 %) is used for climate control through the pre-heaters and re-heaters.

The electric energy is estimated to be somewhat more evenly distributed: 37 % for main power plant (chillers), 25 % for air/water circulation, 20 % for laboratory equipment and 18 % for lighting.

These figures were arrived at by current observations and deductions since no metering of subsystems existed and design data is not realistic. Two factors contributing to this design data inaccuracy are the allowance for high thermal load in the laboratories for vacuum tube instrumentation which was predominant when the facility was constructed, and the continuing conversion of laboratory space to office space.

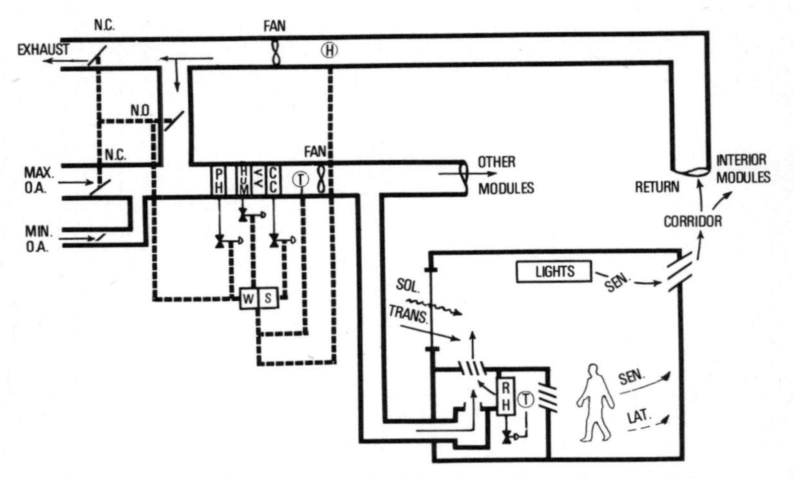

Fig. 3. *Typical Perimeter Modules* (*Terminal reheat induction units*).

Using the energy from the steam and chilled water lines, each building's space is conditioned by a heating ventilating air conditioning system (HVAC) which has been designed to offer maximum flexibility and control. These systems, sometimes as many as 10 or more per building, are of five or six different types; however, over 80% of the space is serviced by only two types:

(1) Low pressure induction reheat units for those modules on the outer building perimeter.
(2) Terminal reheat system for the interior modules.

Figures 3 and 4 are schematics of these systems, respectively. Both of the systems derive their ability to control temperature and humidity through simultaneous heating and cooling, a method inherently wasteful of energy.

For the perimeter modules, outside air is brought in to the zone systems, mixed with recycled air and passed over a preheater (PH) and humidifier (HUM), which may be activated during winter operation, and cooling coil (CC) for controlling maximum dew point which is usually activated throughout the other three seasons. The air is then ducted to individual modules where it is 'jetted' through a slot which induces an additional 15 to 20 % air from the room to be circulated over the reheat coils (RH). Air is exhausted above the module door to

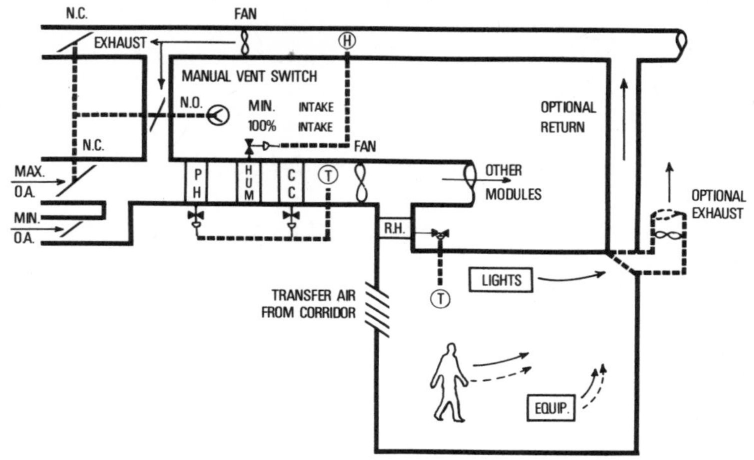

Fig. 4. Typical Interior Modules (Terminal reheat coils).

the corridor, from which it is drawn either into the interior modules or directly to the return system. The pertinent energy controls consist of a thermostat (T) between the cooling coils and supply fan for maintenance of the dew point, usually set at 12·8 °C (55 °F), a thermostat at the reheat coil for maintaining the room temperature at 23·9 ± 1 °C (75 ± 2 °F), and a humidistat (H) in the return duct for maintaining the relative humidity (45 % ± 5 %). The winter/summer switch (W/S) is manually set at prescribed times of the year. Typical loads imposed on this system are solar, air infiltration, conduction through exterior walls, lights, and both sensible and latent gains from people and equipment.

For the interior system, the sequence of events for the supply air is similar to that of the perimeter modules except that the supply air is passed over the reheat coils directly. Many modules have either an exhaust or hood (≃ 500 in total) with a fan venting 100 % of that module's air directly to the outside. In these cases about one third of the total interior module air is taken from the perimeter modules via the corridor. Controls are somewhat simpler in that there is no winter/summer switch option and thus no winter economiser cycle. Loads differ from perimeter modules in that environmental effects are much less significant (except supply air inlet conditions) and internal sensible and latent gains are most important. Predicting the equipment loads is a problem, since they are highly variable from module to module and day to day, thus maximum design conditions are assumed, with reheat supplementing the actual internal load to reach design point.

Immediate-Impact Programme (Operations)

An Energy Task Force was established at NBS to effect maximum energy conservation without curtailing programmes or services, to develop contingency plans to keep the laboratories in operation in the event of severely reduced energy supply, to assist employees with transportation problems, and to educate the staff on the importance of energy conservation. The results of only the first of these tasks are discussed here; reference[1] contains the methodology and results of the other tasks.

The general procedure followed by the Task Force was to appoint sub-working groups in the areas of heating fuel and electricity. The groups attempted to quantify the typical energy usage in the various subsystems, identify energy conservation measures, and through close liaison with Plant Division implement these measures. These measures were limited to operations only, with little or no capital investment for hardware.

The specific measures included:

(1) *Lighting Reductions**—one third of the fluorescent tubes and ballasts were removed from all perimeter modules which still allowed for 614 LX (57 foot candles) of artificial lighting plus any natural lighting through the exterior window.

* It is recognised that in a building with a reheat system lighting reductions usually do not reduce net energy usage because the reheat makes up the load necessary to maintain design conditions. However, the present costs for electricity are approximately 4 times that for heating fuel; thus this is a prudent economic measure.

(2) *Thermostat Adjustments*—a previous schedule of a constant year-round dry bulb of 23·9 °C (75 °F) in all building zones was altered in the office areas to a schedule of 20 °C (68 °F) in winter and 22·2 °C (72 °F) in summer, which are the lower limits of the ASHRAE comfort zones.[2]

(3) *Off-Coil Temperature Adjustments*—in those zones where 50 % relative humidity is not critical, namely the office areas, the cooling coil thermostat was raised a few degrees to 14·4 °C (57 °F), also the summer/winter switch schedule was changed when practical.

(4) *Building and Zone Shutdowns*—the main supply and return fans were turned off during the evenings and weekends in part or all of six of the outlying buildings and the 11-storey administration block.

The results of the first year's implementation are approximately a 12 % reduction in electricity and an 18 % reduction in heating fuel. This is a conservative reduction in that the measures were not completely implemented until the 4th month of the fiscal year (Oct. '73). Referring to Fig. 5, it may be seen that the dashed line represents the total site electricity consumption on a monthly basis for the pre-conservation reference year. The two solid lines represent the consumption since the beginning of the conservation campaign. It

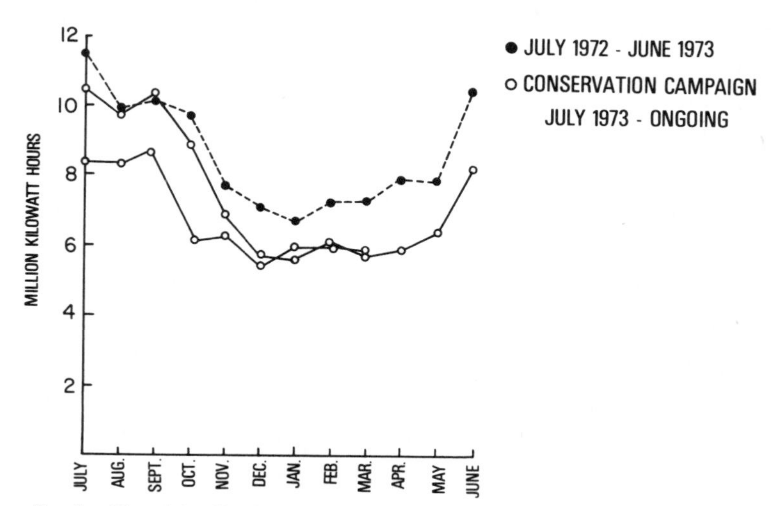

Fig. 5. *Electricity Use From July* 1972 *to March* 1975 (*NBS, Gaithersburg*).

may be noted by the two coincident curves that once the measures were fully implemented (Nov. '73) additional attempts to find significant savings through operations were not successful. Figure 6 is a similar histogram for heating fuel* usage. Again, the pattern of energy savings appears after the measures were fully implemented, with the second year's usage differing only for weather. Figure 7

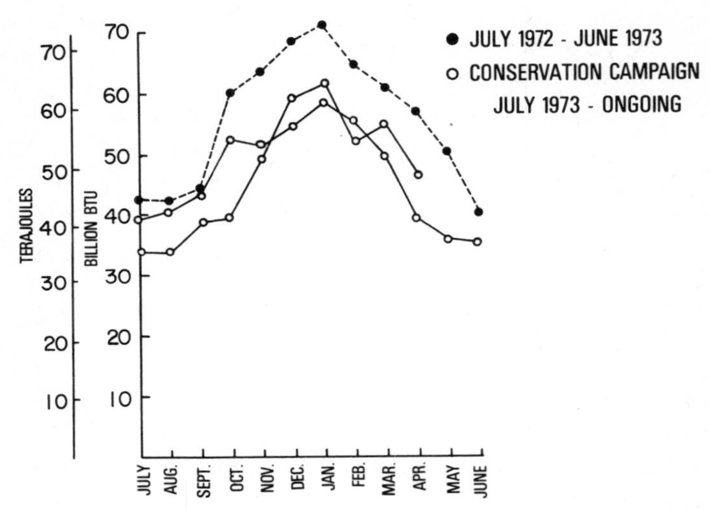

Fig. 6. Heating Fuel Use From July 1972 *to April* 1975 (*NBS, Gaithersburg*).

represents a correlation of heating fuel usage with outdoor dry bulb temperature. Usually it is not possible to correlate energy usage with dry bulb because of the latent load of the air water vapour. However, in a reheat system the air which is treated by the heating fuel has already been cooled to a constant dewpoint in the higher temperature periods so the fuel demand is a function of the outdoor load only through the building walls. The upper curve represents two preconservation years and the lower curve the post conservation period. Thus the heating fuel savings which resulted from the operations conservation measures (without weather considerations) is approximately 25 % of that used during the warmest months and 15 % of that used during the coldest months.

* The fuel quantities were normalised to energy units by assuming 1013 Btu/cu ft for natural gas and 139 000 Btu/gal. for No. 2 oil.

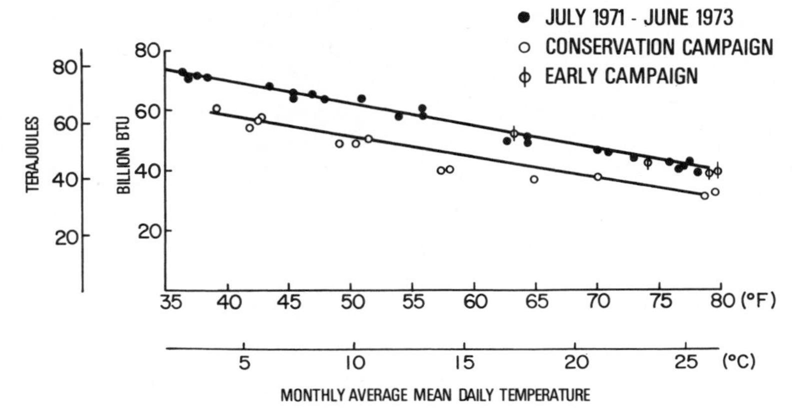

Fig. 7. Heating Fuel Use VS Outdoor Temperature From July 1971 *to April* 1975 *(NBS, Gaithersburg).*

The Long Range Study

Simultaneously with the immediate-impact programme, a long range study programme was initiated through a group of engineers from Plant Division and the Centre for Building Technology. The goal for this effort was to determine which permanent alterations to the building complex would be most energy and cost effective. After much deliberation the topics which were considered to merit detailed study were:

(1) Evaluation of window loads.
(2) Optimisation of year-round inside temperature.
(3) Reduction of ventilation rate.
(4) Evaluation of shutdown of HVAC units.
(5) Optimisation of HVAC control.
(6) Reduction in the number of exhausts.
(7) Optimisation of chiller operation.
(8) Evaluation of exhaust heat recovery units.

The effectiveness of the first five of these topics was approximated through the use of a simulation model which was developed by combining the NBSLD Thermal Load Computer Programme[3] and the HVAC systems portion of the Meriwether Building Systems' Computer Programme. The sixth topic involves considerable occupant cooperation and the seventh and eighth topics were considered to be of secondary effectiveness. Therefore the evaluation of these latter three topics has been delayed.

The simulation model considered the site as a series of buildings with identical shells and HVAC systems subjected to an annual weather cycle constructed from a ten year average for Washington,

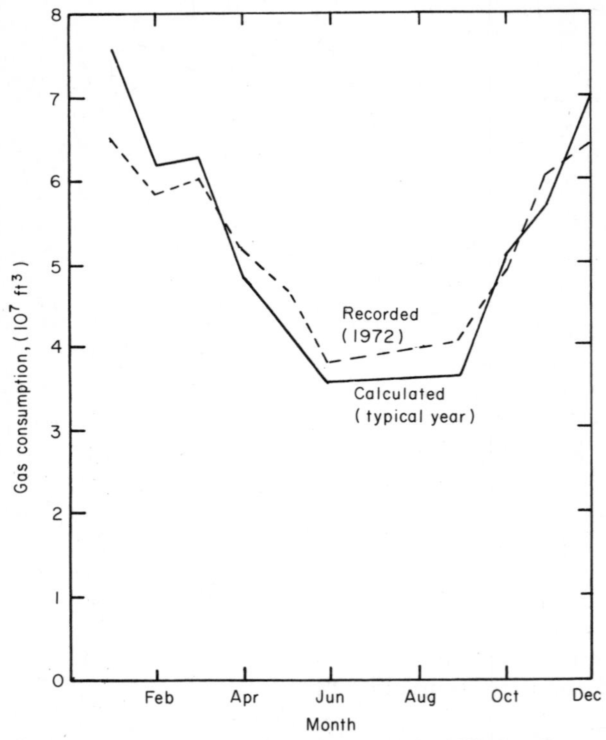

Fig. 8. Comparison of Calculated and Recorded (1972) *Gas Consumption.*

D.C. The validation of the model and the input assumptions were made through a comparison of the heating fuel consumption and chiller electricity consumption, the results of which may be seen in Figs. 8 and 9. Although the measured and predicted consumption results for any given month may differ by varying amounts, the annual results agreed within 7%. The differences between measured and predicted results are in part due to the average annual weather versus the 1972 weather and in part due to the input assumptions for items such as air infiltration or equipment load. It is of course possible to have cancelling errors in the input assumptions in such a way that the results might suggest an accurate simulation; therefore until further submetering data are available the model has been used only to compare relative differences in consumption for the various energy conservation topics if they were applied campus-wide. A summary of results is itemised in Table 1.

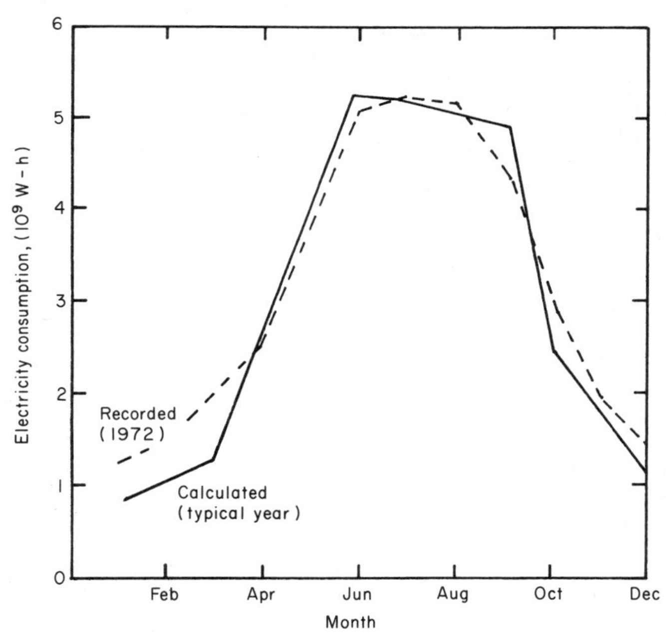

Fig. 9. Comparison of Calculated and Chiller Portion of Recorded (1972) *Electricity Consumption.*

The analysis of the first topic, evaluation of window loads, indicated that if the $3{\cdot}7 \times 10^4$ m^2 (4×10^5 ft^2) of exterior glass (22 % of total wall area) were retrofitted with an additional layer of glass, with an intervening air gap, the net energy savings would be less than 10 %. Although this is not an insignificant amount the relatively high investment required for this operation made other options more attractive. It should be noted that these savings would result from a reduction in load only and that additional savings would be possible if the corresponding decrease in supply air flow rate was made. This additional saving has not yet been estimated.

The optimal indoor temperature setting is of course the same as any reheat system, the lower the better. The magnitude of the savings in heating fuel indicate that it is one of the topics to be considered first for implementation, particularly in view of the fact that it requires no hardware investment but does require considerable labour investment (700 man hours) to reset approximately 2000 thermostats.

TABLE 1
EFFECT OF VARIOUS CONSERVATION MEASURES ON ANNUAL HEATING AND ELECTRIC (CHILLER) ENERGY CONSUMPTIONS

	Annual heating energy			*Annual chiller's electric energy*	
	10^9 *Btu*	10^6 *kWh*	%'72	10^6 *kWh*	%'72
Reference year (1972)	626	183	100	35·6	100
Insulating glass	559	164	95	32·0	90
Temp. set to 25·6°C (78°F) (Interior)	714	209	114	33·3	94
Temp. set to 22·2°C (72°F) (Interior)	548	160	87	35·2	99
20% air reduction	528	155	84	30·5	86
Unoccupied shut down	215	63	34	14·4	40

The 20% air reduction was selected because it was determined to be the amount which could be tolerated in the office spaces that had their lighting load reduced by 33·3%. The 16% savings in heating fuel has been verified by an experiment in several office modules where the reheat water flow has been monitored and the supply air flow varied. However, this reduction in supply air, which is jetted through the nozzle causes a complementary reduction in induced room air which passes over the thermostat sensor. Whether the reduced air system can retain control during extreme weather periods remains to be seen.

The shutting down of additional HVAC units has, by far, the greatest potential for savings of any conservation measure. The results predicted in Table 1 correspond to an operating schedule of only 12 h/day for 5 days/week for the entire campus. Shutting down is defined as turning off the supply and return fans of the air conditioning units but leaving the reheat and chilled water pumps on. The shutting down of the entire campus would not be practical because of the numerous laboratories with exhausts and special environment requirements. In addition, many zones will have a varying schedule due to occupant needs (weekend work, etc.). For these reasons and the fact that there exist more than 150 separate air conditioning zones throughout the campus, it was deemed necessary to focus the study on the HVAC industry's relatively new concept of automatic central control system with a real-time computer optimisation for energy usage.

Automated Central Control System

The use of energy-optimising controls on heating and air conditioning systems is quite new. Therefore, it was decided that a measurement of performance of a variety of control options in a single large building was desirable before selecting the particular control system for the whole site. A full-building experiment, utilising one of the general purpose laboratory buildings, with all ten of its air conditioning zones and energy flows being monitored and with central readout and control being located in the plant, is presently under construction. Two of the perimeter zones, one facing north and one facing south, are having their control systems modified by outdoor enthalpy and supply air temperature reset controls as illustrated by Fig. 10 (compare with Fig. 3). One interior zone is being modified by an indoor schedule for supply air temperature reset and variable volume as shown in Fig. 11 (compare with Fig. 4). The building will be simulated by the analytical model and the measured data taken

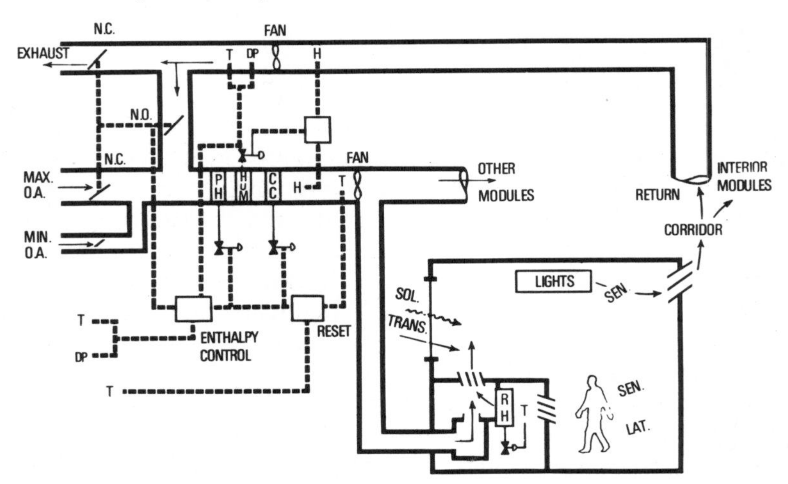

Fig. 10. Modified Perimeter Modules (Terminal Reheat Induction Units).

during three seasons (winter, spring, summer) will be expanded to determine the predicted annual performance of each respective control modification.

The specific control systems selected for evaluation were outdoor enthalpy, supply air temperature reset based on outdoor drybulb, supply air temperature reset based on the drybulb temperature of the indoor module having the greatest cooling demand in that zone, variable air volume, and the already existing economiser cycle.

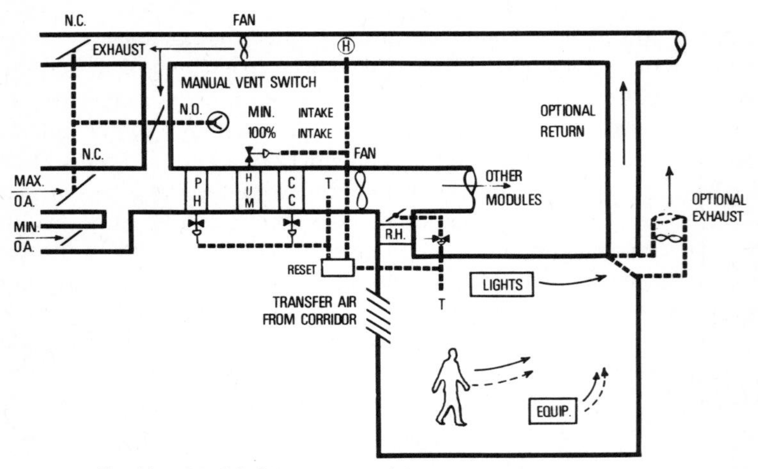

Fig. 11. *Modified Interior Modules* (*Terminal Reheat Units*).

The significance of enthalpy control may be noted by comparing the psychrometric charts in Figs. 12 and 13. In general, the enthalpy control will activate the existing economiser controls as the outdoor enthalpy drops below the return air enthalpy, whereas the present system requires the switch to be activated manually. This is done twice a year based on the monthly average outdoor drybulb temperature. The two triangular regions formed by the outer bounds of the psychrometric chart, the return air enthalpy line, and the return air (RA) drybulb temperature line are the specific zones affected because they involve reversing the inlet air conditions from minimum to maximum and visa versa. The accumulated hours in Washington, D.C. when the outdoor dry bulb temperature is less than the return air dry bulb temperature, and the outdoor air enthalpy is greater than the return air enthalpy are approximately 14% of the total for the year.

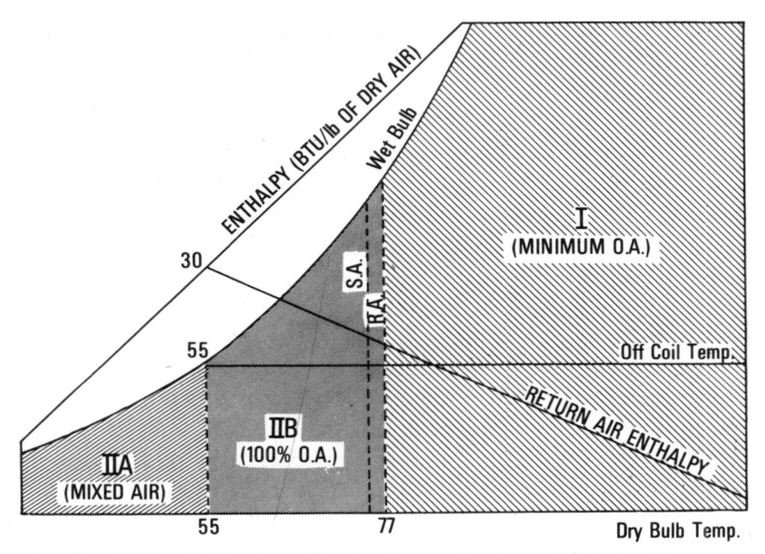

Fig. 12. NBS—Gaithersburg Site Psychrometric Chart of Existing Controls.

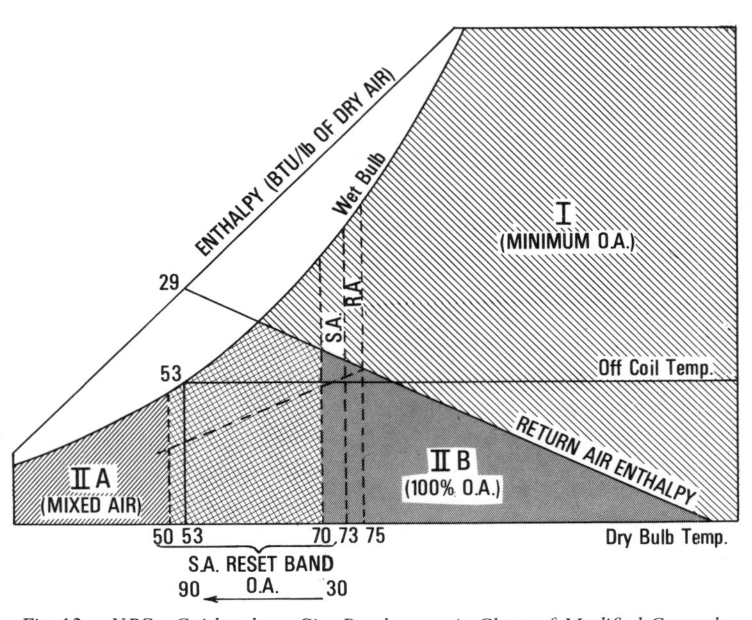

Fig. 13. NBS—Gaithersburg Site Psychrometric Chart of Modified Controls.

On the other hand, the accumulated hours when the outdoor dry bulb temperature is higher than the return air dry bulb temperature and the outdoor air enthalpy is less than the return air enthalpy (zone IIB on Fig. 13) comprise less than 2% of the year. This relationship in total hours of occurrence caused us to include the enthalpy control for further study and to reject an additional mixed air control for the IIB zone (Fig. 13) which would have optimised the return air and inlet air to give the desired off-coil temperature directly.

The perimeter module air temperature reset control, with a schedule based on outdoor dry bulb, will automatically reset the off-coil thermostat's control point as high as possible while still maintaining a dewpoint that will allow the room conditions to be at or below the desired relative humidity level. This provides a continual increase in the control point, as the outdoor air temperature decreases, and thus a decrease in the demand for reheat as well as cooling. The operational band possible for this control is shown in Fig. 13; actual band limits have yet to be determined for the NBS site. Also, whether it is desirable to operate this control simultaneously with the enthalpy control and the minimum drybulb range over which this system might be activated is yet to be determined. The interior air temperature reset will operate on the same principle except that it will respond to a schedule for the interior module with the greatest cooling load. It also requires a direct humidstat override since it is possible that a given laboratory could have an excessive humidity load.

The variable air volume system will be created by installing servo-motors on the existing supply air ducts for each module. Upon demand for heating, the pneumatic control signal from the module's thermostat will first reduce the supply air to a set minimum and then open the reheat valve.

Each of these systems will be evaluated for energy savings and dollar investment requirements for possible application to other zones of the pilot building and later to other buildings in the NBS laboratory complex. The data from the pilot study and the analytical model will be used to calculate potential savings if the best combination of the control systems were applied to the entire pilot building and to all ten buildings on the campus for which such a system now appears to be feasible.

It is apparent from the above discussion that this report is of an interim nature describing an ongoing experiment in system control. The collection of data in the pilot building is scheduled for initiation this winter, and it is anticipated that the experimental systems can be fully evaluated by early fall of 1976. If the experimental systems are proved to be beneficial both in energy savings and investment cost, broader application is planned for the winter of 1976–77.

The system control concepts being studied, if successful, will have application to the design of new buildings as well as to the modification of existing buildings which utilise similar reheat systems.

REFERENCES

1. Hoffman, J. D., *Energy Conservation at the NBS Laboratories*, NBSIR-539, National Bureau of Standards, Washington, D.C.
2. *ASHRAE Handbook of Fundamentals*, (1972). American Society of Heating, Refrigeration and Air Conditioning Engineers, New York.
3. Kusuda, T., *Computer Program for Heating and Cooling Loads in Buildings*, NBSIR-574, National Bureau of Standards, Washington, D.C.

Discussion

N. F. Bradshaw (*Haden Carrier Ltd*): I would like to ask the American speakers if they can explain the motivation in the USA to energy conservation. A number of us who went to Boston in June 1975 are under the impression that Americans will put effort into trying to save energy, which is a fraction of their costs. American Government organisations will sponsor studies energetically, American people will accept a reduction in the speed of their motor cars. Is it perhaps that you do not like the idea of being dependent on imports of oil?

J. F. Barnaby (*Connecticut General Life Insurance Company*): You pose a very difficult question to answer—I can give you a hundred answers. I think the most honest answer is the one you hit—we are tremendously independent—we are tremendously proud and I think we resented deeply the Arab oil embargo.

P. R. Achenbach: Mr. Bradshaw by suggesting the USA was ahead of other countries in carrying out an energy conservation programme may have paid us a compliment we do not deserve. Our discussions in the USA cover the same goals, methodologies and problem-solving efforts as those being developed here.

In identifying earlier the incentives for energy conservation in buildings, conventional terms were used such as exhortation, education, raising prices, government subsidy, building regulations and rationing. As I listened it seemed to me that the motivations for action could more accurately be described as profit, regulatory authority, deprivation, and concern for present and future human welfare. I believe that we are trying to implement an energy conservation programme for buildings in both the USA and the UK based on a concern for present and future human welfare when this motivation is probably the weakest motivation on the present scene.

Another worthwhile way to look at energy conservation is to identify the groups which have the levers to make something happen. Of course, the suppliers of fuel and energy, both foreign and domestic, can reduce energy usage and cause deprivation by decreasing the supply. The design professionals represented at this conference have the technical knowledge to design buildings and systems that require less energy, but many professionals feel that they must have higher fees to provide this service. Mr. Dubin suggested in presenting his paper that energy conservation design of buildings can often be carried out profitably at present fee levels, but that capable professionals can obtain higher fees in return for cost-effective energy conservation proposals. I believe that Federal government officials have a responsibility to provide leadership in promoting short and long-term public welfare with respect to energy conservation even if it puts their popularity and job security in some jeopardy. Of course, the public can exert some influence on energy policy by the choices made at the polls and by refusing to occupy buildings requiring excessive energy to serve them.

There have been substantial differences of opinion in the USA about the best combination of incentives to reduce our energy requirements, although most people pay lip service to the need for conservation. There is a considerable thrust in the USA toward the development of standards for energy conservation in both new and old buildings that could be used as building regulations if supply, economic or national security conditions warranted such action in the future.

F. S. Dubin (*Dubin-Mindell-Bloome Associates P.C.*): We are becoming increasingly concerned in the USA with the environment. For instance, the motivation for a recent study in Long Island was the fact that a power plant site survey disclosed that Long Island was the most suitable site for power plants in the whole of New York State. The people of Long Island could visualise wall-to-wall power plants all round their beautiful island. That became a very powerful motive force to examine alternatives.

Some of the engineers have been really taking a lead—they have suggested schemes to the government—they have suggested many of the studies in unsolicited proposals. The GSA energy conservation demonstration building now under construction in New Hampshire came about through the impetus of engineers asking 'isn't there already enough paper, enough discussion?' 'isn't it time to build a demonstration building?'. At that time Commissioner Sampson of GSA agreed, and announced that they would build a demonstration building—in fact they did two. Motivations come from many different directions, however I do not think there is any tremendous mass movement. Most of the actions have come about by the effort of a few dedicated people that have influenced others. A few can really make things happen, if they really put their mind to it. There are many dedicated professionals and we have very many people, at least at some levels in our government, who are truly concerned with the energy supply-demand picture and they are implementing ideas into programmes. I doubt if the action being taken in the USA is all that much more than you are doing in the UK. Because we are very much larger and have a larger population, it seems that our efforts are more extensive but on a *per capita* basis, I am not so sure that we are far ahead.

5

Swiss Examples of Energy Saving in Existing Buildings

MAINI CARATSCH

Synopsis

General situation.

Fuel saving in heating installations.

Energy saving in a 24 storey office building, by modification of windows and modifications of service.

Comparison of energy consumption of different air-conditioning systems.

Introduction

Before explaining the measures taken to save energy in existing buildings, it is important to understand the context into which energy saving has been injected.

Most of Swiss industry and the majority of the Swiss population are situated in the central part of Switzerland between the Jura and the Alps on a hilly plane about 400 m above sea level. The climate in this area is such that heating is required during about seven months of the year. The design outside temperature for Zurich is −11 °C although, of course, lower temperatures than this are experienced during the heating season. The heat requirement is estimated for 3700 degree-days and in more mountainous regions this could go up to 6000 degree-days or more.

Summer time is rather short and the outside temperature reaches 30 °C on only a few days a year with night-time temperatures 12° to 16 °C or less.

Swiss buildings are designed and built to meet these climatic conditions and now that wood is too expensive to use as building material, concrete and brick are the main materials. In the 1930s a 380 mm wall with a U Value 1·16—1·5 W/m^2 °C, was the rule. The insulation was achieved by a 30 mm layer of insulation material sandwiched between two brick walls each 120 mm thick. In office buildings the curtain walls were made of metal panels filled with insulation U Value 0·9—1·00 W/m^2 °C. Windows were treated with special care: there was no building without double glazing.

Increasingly insulating glazing is used for offices and also for blocks of flats, even for single family houses and bungalows. Aluminium frames increase the tightness of the window but are still seldom used in private houses because of the high cost.

In the 1930s the main fuels were coal and coke. Office buildings and some dwellings had central heating but many flats had single stoves burning wood and coke. Since the 1950s all new buildings have been centrally heated, generally using oil and many existing coal boilers have been converted to burn oil. Oil brought a number of advantages, in particular, it was now possible to provide automatic controls to ensure the correct supply water temperature, previously over-heating had not been uncommon with occupants controlling the temperature by opening and closing windows.

In Switzerland today about 85 % of all buildings are equipped with central heating and out of these, 90 % are oil-fired. Gas is very rarely used for heating in Switzerland which has no natural gas and no coal deposits.

The very low cost of oil persuaded people to heat domestic hot water with an oil-fired boiler instead of the electricity which had been used previously. Combined boilers for central heating and domestic hot water supply became common practice with the domestic hot water being supplied by the boiler even during the summer.

Mention must also be made of the air-conditioning installations which have increasingly become standard equipment in administrative and office buildings, hospitals, laboratories, shopping centres, etc. There are a number of reasons for this air-conditioning boom and these may be summarised as follows:

The building: lightweight external envelope with very low heat capacity and high percentage of glass produces considerable summer heat gain.

Outside air: increasing air pollution (dust, smoke, exhaust gases) make thorough air filtration increasingly desirable.

Traffic noise: at current levels, working near an open window has become intolerable.

Utilisation of building space: occupancy can be increased and full use made of core areas.

The most common air-conditioning systems are the induction system, often using free-cooling and the dual duct system with the possibility of using fresh air for cooling. New buildings are equipped with some kind of heat exchanger for heat recovery, commonly the Econovent-Wheel.

Systems incorporating reheating coils for various zones are only applied where exact humidity control is required. They are normally considered uneconomical for comfort air-conditioning.

This short description will indicate that buildings in Switzerland are well built. Considerable consideration has always been given to insulation, including the effect of windows with the result that heating and air-conditioning insulations have reached high standards of efficiency and automatic control.

It is important also to realise that in a democracy restrictions are difficult to enforce except in cases of emergency. The environment problem is soon forgotten and even constant reminders will only have limited success. The most effective impetus for energy conservation is energy cost. High oil prices will result in considerable effort being expended in finding ways and means of reducing fuel bills.

Combustion Efficiency

Approximately two years ago, following pressure from the environmental lobby, the Swiss government made a regulation that the combustion efficiency of every boiler must be checked annually. Although this was intended as an attempt at pollution control, combustion efficiency also has implications in terms of energy conservation. Fifty-three percent of all fuel imported is used for heating buildings ($7 \cdot 3 \times 10^6$ tons). A general increase in combustion efficiency of 2% would save 150 000 tons. Such an increase should not be difficult to achieve. In 1974 fuel consumption was reduced by 10% compared to the previous year, although this was partly due to a mild winter it also resulted from better control and servicing of heating equipment.

Reduction of Fuel Consumption

A number of steps have been taken in Switzerland as a whole to reduce fuel consumption, for example,

(a) Room temperatures have been reduced from 22° to 20°C.
(b) Windows have been made more weatherproof.
(c) There has been an increase in the use of individual radiator control valves.

Using figures for Zurich, the reduction in room temperature alone will give a 12% fuel saving (based on 4200 degree days for 22°C and 3700 degree days for 20°C room temperature).

The energy saving is obvious but the important question is whether 20°C is an acceptable room temperature for comfort. The answer is, I believe, that if 20°C is the average of the air temperature and the temperature of the surrounding walls, then it is adequate. Such a situation can only be achieved if the outside walls are well insulated and adjacent rooms are heated and, perhaps most important of all, windows are double glazed. It is important to avoid large temperature differences between different walls as this would create a one-sided radiation condition and considerable discomfort to occupants.

Air infiltration can easily amount to half the total heat loss. If a wooden-framed window is weatherproofed to reach the standard of a metal-framed window, this would result in Switzerland in a saving on heat losses of 8% where window area represented 20% of exterior walls.

Radiator control valves on each radiator bring energy savings because the heat supply can be controlled to the individual requirements of the room, for example, additional solar heat gain, light, people or electrical appliances such as television sets, will result in activation of the radiator valve to reduce the water supply to the radiator. Some manufacturers have claimed that such valves can save up to 20% of the fuel bill.

With solar gain alone I estimate 7% of the fuel requirement could be supplied from the sun resulting in the same reduction in fuel consumption (calculation based on 620 sunshine hours and a total of 8500 degree hours).

These three changes require very little money. Weatherproofing of windows is generally effected by 'do-it-yourself' and supply and fixing of automatic radiator valves cost about 70 Sw Fr. each.

Such energy conservation measures have been made in office buildings as well as houses and flats in Switzerland by economic pressures alone. In France automatic radiator control valves are even recommended by the authorities. Automatic radiator control valves are particularly applicable in terms of energy saving to installations where no other automatic control has been installed. They will also result in higher energy saving in hilly areas where the hours of sunshine are greatly in excess of those in Zurich where the climate tends to be cloudy.

Domestic Hot Water Supply

The increase in the use of a single oil-fired boiler for heating and domestic hot water supply has already been indicated, the domestic hot water being heated by gravity circulation through special waterways in the boiler. The domestic hot water is heated by the oil-fired boiler even in summer, the oil burner being used on very intermittent operation. Using a large boiler to supply a small summertime load is very inefficient because of additional heat loss to the flue, high transmission losses and low combustion efficiency. An overall efficiency of 20% would be a reasonable achievement. The efficiency can be increased to 50% by changing the control system such that the burner is switched on only three times a day with the water content of the boiler effectively used as a storage tank and the use of bigger differences of water temperature. Naturally, it will be preferable to have a separate boiler with its own circulation pump and a capacity equivalent to a day's requirement of domestic hot water. A problem here is the additional investment required for such a change once you have the combined boiler. An increase from 20% to 50% efficiency in warm water supply, it is estimated, would save about 630 kg of oil per annum for a family of four using water at 45 °C.

Solar Energy

Solar energy has been rediscovered in the last two years and considerable effort has been expended on consideration of its use for heating purposes, indeed, a special society to propogate the use of solar energy in buildings has been established. A few installations are already operating, some with heat pumps, some only to heat domestic hot water.

The solar heat gain in Zurich is about 3·6 GJ/m^2 annually although 70% of this is received during the summer months. The most reasonable use of solar energy would seem to be to heat water in a storage tank for domestic hot water service. For the four person family mentioned above this could mean a saving of 1050 kg of fuel (the total oil used for water heating during the summer). It is interesting to examine the possible cost; with 8 hours of diffuse solar radiation intensity about 200 W/m^2, a collector surface area of 10 m^2 would be required to heat 400 litre of water from 10° to 45 °C at an estimated cost of 400 Sw Fr./m^2. Installed, the total cost would be 4000 Sw Fr. but the saving in fuel would only be 340 Sw Fr. This rough assessment indicates that even if the radiation intensity were higher and some use of the collector were made during the heating season, at the moment a solar collector is not financially economic. Being developed are new collectors which are likely to be cheaper in price but, unfortunately, the rather low solar intensity and our rainy and cloudy days will not substantially change.

Air-conditioning

Reduction of room temperature, relative humidity, fresh air ventilation rate and running time have all been applied to installations in order to save energy.

Many clients have asked us to study the installation of heat recovery devices. We have found that heat recovery, in existing installations, would be very difficult to achieve owing to lack of space and it is our experience that clients soon lose interest once they know what the cost would be. The scope in the installations we examined was very small because of the high standards of efficiency and maintenance already in being.

Energy Conservation in a Multi-storey Office Building in Zurich

The following section deals with a specific energy saving exercise carried out in an office used by Sulzer Brothers. The building is 30 m by 30 m square and contains 24 storeys. Up to the 22nd floor the windows are shielded by exterior Venetian blinds. The 23rd and 24th floors have larger windows and blinds on the interior as exterior blinds were not permitted. The whole building is air-conditioned using a dual duct system and air distribution is through a perforated false ceiling from 230 mixing boxes. About 30 m^3/h of chilled air is blown along each running metre of window in summer. This is replaced by dry heated air in winter. The intention is to remove part of the heat irradiated by the Venetian blinds in summer and to prevent a cold down draught in winter. The conditions proved to be comfortable on all floors except the two upper floors. The solar heat gain to these floors was much higher and the inner Venetian blinds reached a temperature of 40 °C and more during the sunny periods. The offices were uncomfortable and even with twelve air changes per hour the room temperature would reach 27 °C. It was decided to equip the two upper floors with a third window, the Venetian blinds being retained between the second and third pane to produce a triple glazed window. The air extract was arranged to exhaust between the two inner panes of the window.

This solution offered three advantages:

(1) Reduction in the temperature of the inner leaf of glass to about 2 °C below the room air temperature means that personnel are comfortable even close to the window with consequent better utilisation of the available floor space.
(2) The highest summertime glass surface temperature measured was 31 °C, a reduction of some 10° from the value previously. The temperature is still however higher than in the rooms using the outside blinds.
(3) The air change rate could be reduced to six air changes per hour with a consequent reduction in energy use.

Tables 1, 2 and 3 summarise the calculated estimate of the yearly energy consumption for the 23rd and 24th floors.

This case is, I think, especially interesting because the solution using an exhaust window could be applied to many existing buildings especially buildings only equipped with single glazing. Not only would energy be saved but comfort conditions improved and the air-conditioning installation required would be smaller than would be the case if Venetian blinds were used alone. Also, the floor space could be

TABLE 1
ROOM CONDITION AND SIZES

Outside conditions	according to the statistical mean temperature of each month as well as the statistical monthly sunshine hours.		
Inside conditions	Winter	Transition period	Summer
Temperature (working days)	22 °C	22 °C	24 °C
Temperature (nights and holidays)	15 °C	15 °C	
Working hours		07·00–19·00	
Working days		253 days/year	
Window area per module		$1{\cdot}75 \times 2{\cdot}60 = 4{\cdot}55\,m^2$	

used right up to the window; no 'safety gap' is required because of draughts or irradiation.

In the building described the cost of each single glazed window was particularly high at around 1500 Sw Fr. because of the luxury standards of finish demanded. Similar results would have been obtained by using even a transparent, but an airtight, curtain as the third window. It is estimated that the costs of such a curtain would be recovered within two years by the energy saved.

The alterations to the two upper floors were primarily to improve comfort conditions but previously the possibilities of further fuel savings were investigated. These are detailed below along with the savings that have resulted. These figures are applicable to the whole building.

TABLE 2
WINDOW DETAILS

Existing window (double glazed with inside Venetian blinds)	
U (overall thermal transmittance)	$= 3{\cdot}50\ W/m^2,\ ^\circ C$
Solar reduction factor	= 0·44 (for global irradiation)
Modified window (triple glazed exhaust air window with Venetian blinds between inner panes)	
U (equivalent overall thermal transmittance) value:	
in case of exhausting air through window	$= 0{\cdot}40\ W/m^2,\ ^\circ C$
in case without exhausting air through window	$= 1{\cdot}98\ W/m^2,\ ^\circ C$
Solar reduction factor	= 0·22 (for global irradiation)

Besides the improvement of the comfort conditions a considerable saving of energy was the result of our modifications.

Table 3
THE ANNUAL ENERGY REQUIREMENT FOR ONE WINDOW

(a)	Heating energy requirement (Transmission loss)		Double glazed with inside Venetian blinds	Triple glazed exhaust air window with Venetian blinds between the inner panes
(i)	During working hours day time (07.00–19.00)	GJ/a	2·20	0·251
(ii)	During non working hours day time (07.00–19.00)	GJ/a	0·41	0·230
(iii)	During non working hours night time (19.00–07.00)	GJ/a	1·90	1·076
	Total yearly heating energy requirement	GJ/a	4·51	1·557

(b)	Cooling energy requirement		West	South	East	North
(i)	Double glazed with inside Venetian blinds. Transmission + Solar irradiation	GJ/a	2·53	4	2·60	0·02
	Electrical energy requirement for supply and return air fans	GJ/a	2·86	3·20	2·86	1·33
	Total electrical energy requirement	GJ/a	5·39	7·20	5·46	1·35
(ii)	Triple glazed exhaust air window with Venetian blinds between inner panes. Transmission + Solar irradiation	GJ/a	1·55	2·20	1·58	0·005
	Electrical energy requirement for supply and return air fans	GJ/a	1·66	1·76	1·66	1·33
	Total electrical energy requirement	GJ/a	3·21	3·96	3·24	1·335

Fresh Air Damper

The control of the fresh air damper was changed. Previously a dew point thermostat had controlled the preheater valve, closing the fresh air damper and opening the return air damper only if the preheater could no longer maintain the required dew point. As the preheater was oversized, the system was operating for almost the whole of the winter period on 100% fresh air. In the new system the amount of fresh air was automatically controlled according to the outside temperature. This has resulted in a saving of 120 000 kg of oil or 45 000 Sw Fr.

TABLE 4
TOTAL ENERGY SAVINGS FOR BOTH FLOORS DUE TO WINDOW MODIFICATION

No. of windows 4 × 16 = 64 windows

Energy cost: Heating = 10.00 Sw Fr./GJ
Cooling = 14.00 Sw Fr./GJ
Electric = 28.00 Sw Fr./GJ

		Heating energy	*Cooling energy*	*Electrical energy*
Existing window	GJ/a	577·0	293·0	328·0
Modified window	GJ/a	200·0	171·0	205·0
Energy saved due to modification	GJ/a	377·0	122·0	123·0
Annual energy cost savings for 23rd and 24th floors	Sw Fr.	3770·0	1708·0	3444·0
Annual energy cost savings per m^2 office floor area	Sw Fr.	·294	1·33	2·70

Reduction of Humidity

The humidity in the room was reduced in winter from 45% r.H to 35% r.H resulting in a saving of 26 000 kg of oil or 9500 Sw Fr. The reduction of humidity to this extent is not harmful either for health or comfort. The importance of humidity has been exaggerated over the last few years and it seems that with a clean, well filtered, dust free atmosphere, the humidity can safely be reduced to 35% r.H.

Reduction of Plant Running Time

Another area for energy conservation was the electricity used in operating the air-conditioning plant. Cutting down the running time saved 8% of the electrical energy *i.e.* 340 GJ = 10 600 Sw Fr. This was done by cutting two hours a day from the running time. The plant is now used between 06.00 h and 18.00 h whereas previously it had been used from 05.00 h to 19.00 h. There is just one exception to this, if the outside temperature falls below −5 °C the fans are kept running all night with a minimal fresh air supply to prevent ingress of cold outside air and too great a drop of temperature in the offices.

Lighting

Almost half the electrical energy used in the building was for lighting. With stricter discipline insisting that artificial lights were kept on only when required, the consumption dropped from 5090 GJ to 4260 GJ *i.e.* 8 %.

Total Energy Savings

Table 5 indicates the total energy consumption savings for the whole building resulting from the energy conservation measures described above. It must be borne in mind that unlike the introduction of the

TABLE 5

TOTAL ENERGY SAVINGS DUE TO ENERGY CONSERVATION MEASURES

Fuel	1973	705 000 kg.	258 000 Sw Fr.	100 %
Saving due to changing of fresh air damper control		120 000 kg.	44 000 Sw Fr.	17 %
Saving due to lower humidity		45 000 kg.	16 000 Sw Fr.	6 %
	1974	540 000 kg.	198 000 Sw Fr.	77 %
Electrical energy	1973			
Refrigeration		830 GJ	23 296 Sw Fr.	
Pumps		275 GJ	7 726 Sw Fr.	
Air-conditioning Plant		5 060 GJ	142 760 Sw Fr.	
Light		5 090 GJ	143 000 Sw Fr.	
		11 255 GJ	316 782 Sw Fr.	100 %
Saving due to shorter running time		340 GJ	10 600 Sw Fr.	
Saving light		470 GJ	14 600 Sw Fr.	
	1974	10 445 GJ	291 582 Sw Fr.	92 %

third window, these savings have been made with virtually no additional capital cost. The saving in oil is more significant than the saving in electrical energy.

The installation of the third window system has only been completed during the summer of 1975 so that measured results are not yet available.

Energy Conservation in Europe

As far as we are aware in Germany and France, as in Switzerland, existing installations have not been changed to improve energy use. Energy saving measures have been primarily concerned with good-housekeeping and running and maintaining the plant well and efficiently. As in Switzerland, the following areas have been investigated and used:

(a) reduction of room temperature
(b) reduction of humidity
(c) reduction of running time
(d) reduction of light intensity
(e) reduction of outside air during extremes of cold and heat

The situation is very different, however, in new buildings where the question of energy consumption is often a primary one, even before capital cost and the only systems which have a chance of being used today are those which completely avoid the parallel operation of cooling and heating plant which can lead to so much waste of energy.

We have made elaborate studies and calculations which are summarised for an office building in Table 6. The calculations were based on exterior Venetian blinds, with the exception of the case of the exhaust window where the blinds were between the two inner panes. The cooling load in this case was slightly greater than in the others except for the two-pipe induction system. The advantages inherent in having protected inside blinds and provision of comfort conditions during winter will make the exhaust windows one of the most economic systems in the future.

TABLE 6
ENERGY CONSUMPTION FOR DIFFERENT AIR-CONDITIONING SYSTEMS

System	*Consumption* GJ/a		
	Heating	*Refr.*	*Fans*
4 pipes induction	1560	280	180
2 pipes induction	2080	760	180
Exhaust window with single duct	1540	460	620
Dual duct with special window air	2460	360	700
Var. vol. with controlled rad. heating	1540	320	430

In terms of economics, the systems where reheating is avoided are best, therefore, the two-pipe change-over system which is based on reheating and recooling, will always be worse than the four-pipe system where cooling and heating are completely separated. The dual duct system figures are calculated based on the provision of window air controlled by an outside thermostat. In this system heating at the window and cooling in the room at the same time cannot be avoided unless the window air temperature can be reduced by some other means.

It is essential that in the future, architects, building services engineers and builders alike, work together to achieve good heating and air-conditioning at a reasonable price and with acceptable low energy consumption.

Discussion

S. Mulcahy (*Steenson Varming Mulcahy & Partners*): As a matter of technical interest, at University College London, in the Environmental

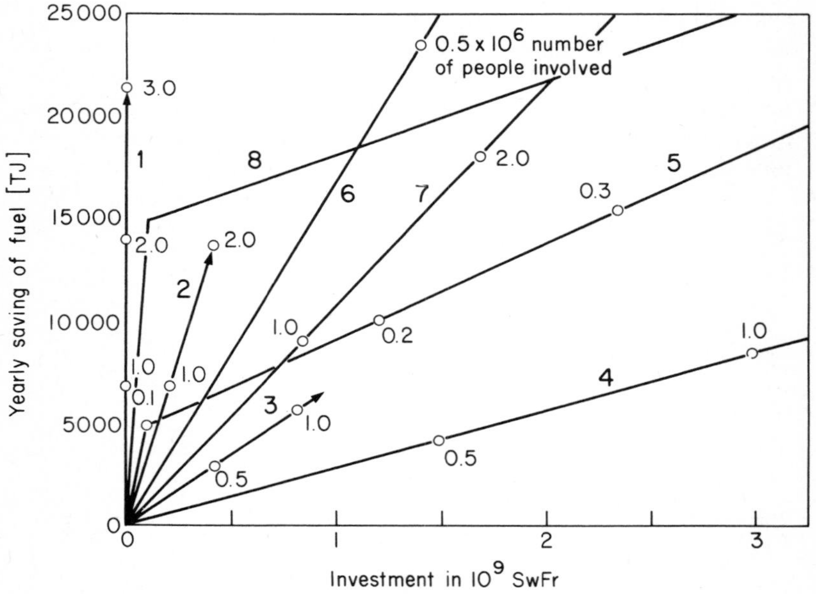

Fig. A. Fuel savings for a specific investment

Services Department, we are experimenting with a low-technology version of Mr. Caratsch's ventilated window by simply having an operable double up and down sash window with a reversible fan which avails of the room heat-gains in winter and rejects them at high level in summer.

B. Marsh (*John Laing Design Associates Limited*) (*wrote*): Delegates appeared at times to be expecting some mystical guiding light to send them home with evangelistic fervour. This, of course, was not to be but, in my opinion, something very near to this was provided by Mr. Caratsch. The slide which showed a number of graphs of energy savings against cost expenditure, and in a nutshell gave in numerical order of benefit, the various options open to the delegates attending. I know the data was on a Swiss base but I doubt if British energy/cost considerations would vary very much. I hope the drawing can be reproduced in the proceedings.

M. Caratsch (*wrote*): The graphs are shown in Fig. A the curves show the yearly saving in fuel for a particular investment for the following options:

1. Reduction of room temperature from 22 to 20 °C (without any investment).
2. Improvement of automatic temperature control.
3. Better thermal insulation on all new buildings to be built during the next 10 years.
4. Improvement of thermal insulation on existing buildings.
5. Electric heating.
6. District heating using waste heat of atomic reactors.
7. Solar energy.
8. Heating by Natural Gas.

The numbers adjacent to the points on the curves $\times 10^6$ indicate the number of people.

6

Energy Economy in Government Buildings

H. P. JOHNSTON and A. C. GRONHAUG

Synopsis

This paper describes the work undertaken by the Property Services Agency of the Department of the Environment on energy conservation and energy management since 1972.

It covers the Agency's energy management policy and procedures, including operation and maintenance of plant and installations, fuel monitoring, economic assessment of utility agreements, power factor improvement of electrical installations, methods for limiting electrical maximum demand, and publicity aimed at the more economical use of buildings and services. Details are given of investigations and current work and views on electrical installations, space heating of buildings, catering installations and the design process for new buildings.

Introduction

The responsibility for most buildings within the Central Government estate falls to the Property Services Agency of the Department of the Environment. A comprehensive estate management service is provided, which includes land and property acquisition, disposal and rental, in addition to design, construction, equipment, furnishing, operation and maintenance activities in support of the wide range of buildings and facilities required by Government Departments and the Armed Forces, at home and overseas.

To meet these obligations, the PSA employs about 50 000 people, including up to 30 000 industrial staff and 11 000 Professional and Technical staff. These are deployed in Headquarter offices in London, Chessington and Croydon; and at nine other major centres in the UK on a territorial basis. Each territorial centre is further divided and subdivided into areas and depots. Similar arrangements, though to a generally smaller scale apply overseas.

Total annual expenditure is of the order of £900 million at current prices, comprising some £400 million for new works and land; £230 million for maintenance (including about £100 million spent on engineering plant and installations); and £270 million on Supplies services, rents, salaries and fuel. Fuel expenditure alone in 1974/75 was approximately £25 million in the civil estate; plus approximately £60 million in the Defence estate, the latter being paid for separately from the Defence vote. Further analysis of the fuel costs into the average amounts paid for coal, oil, gas and electricity across the board, shows that electricity is the most significant single element, alone accounting for one-third of the total energy cost.

In the period 1964–1975, the fuel bill for Government buildings trebled, primarily as a result of price increases. But while the increases in the last two years have been particularly heavy, the upward trend was apparent earlier and had already led to a number of new initiatives aimed at improving performance and reducing the rate of cost increase, before the recent crisis broke. The PSA was therefore in a relatively strong position to respond quickly to renewed calls for national energy savings, once the full extent of the new situation became apparent.

From this starting point, a revised and easily implemented policy was evolved, of which the main features comprise:

a declared aim to achieve a defined savings target;
a restated commitment to uphold nationally agreed standards;
a requirement for all measures implemented to be economic in terms both of costs in use and of staff resources;
a decision to operate the conservation programme so far as practicable, on a self financing basis, using the actual savings achieved in one year, to determine the size of the capital investment to be provided in the next.

The savings target chosen has been set to achieve an overall reduction of 30% in the annual fuel consumption, over a five year period from 1974. This target may prove eventually to be impracticably high, but was set at this level in order to highlight the importance attached to energy conservation in its own right; to encourage inventiveness and initiative in the approach to energy conservation, for both existing and new buildings; and to provide sufficient leeway to ensure that as further energy price rises occur, any marginal schemes which may then become economic, can be kept in mind and re-assessed.

Existing Organisation and Procedures

Since the new energy conservation policy adopted is evolutionary, not revolutionary in concept, it is worth examining briefly the main features of the existing organisation and procedures, which have proved to be of greatest benefit.

Foremost among these is the system of Planned Preventative Maintenance adopted. Full details of this system are described elsewhere,[11] and it is not intended to cover the same ground here. Suffice it to say therefore that the system has been developed over thirty years and is now well proven; that it establishes performance criteria and standardises routines covering operation as well as maintenance requirements, for all the most important and sensitive items of M and E plant and equipment; that it operates flexibly by delegating responsibility to areas and depots, both to vary the frequency of tasks and to introduce any local routines required to suit the particular needs of individual circumstances; that it includes the provision of feedback to the centre, so that significant failures or successes whether of design or maintenance, can be monitored and recorded; and that linked to it (though separately managed) is an M and E training school at which both operatives and engineers can be taught and kept up-to-date in the correct methods, procedures and practices.

While this planned maintenance system may not suit other organisations of different sizes having different problems and priorities, the important point is that without a sound maintenance and operation base, it would be impossible to judge the efficacy of energy conservation measures with any degree of assurance. Some system performing an equivalent function and similarly establishing maintenance and performance norms is therefore an essential prerequisite to any sound energy saving policy.

As an addition to the planned maintenance organisation, a small (five man) Fuel Economy Unit was formed in 1972 to meet the needs of a centralised approach to the efficient and economical use of all fuels and to explore and exploit possible ways of controlling and reducing expenditure. The Unit is fully accountable in its own right and has from the outset had to pay its own way. The Unit is linked to the Territorial Organisation through Fuel Economy Officers appointed in each of the main local centres, thus providing a rapid, two-way, communication channel on energy matters throughout the organisation.

The Fuel Economy Unit initially concentrated its activities in the heating systems control field. It was and remains responsible for the optimum start control programme and in support of this activity has provided courses and lectures on the subject, both for the benefit of PSA staff and for those in other organisations. This in turn led the Unit into other areas of publicity and especially publicity aimed at users, to create a climate of waste avoidance by voluntary means. The optimum start programme has been particularly valuable because as energy prices have escalated, the return on investment has been exceptionally rapid and it has been possible to demonstrate that further measures can be funded from money which otherwise would have had to be set aside to meet fuel bills. In short, this work has effectively helped to prime the pump, in support of the enlarged conservation programme now under way.

On a different level, it is also worth mentioning the specific financial procedures which are used both to decide on the best choice of fuel in particular cases and to assess the merits, or otherwise, of additional capital investment in order to produce longer term overall savings in operating and maintenance costs. These procedures follow the conventional discounted cash flow route, using test discount rates agreed with the Treasury and relative rates of change in fuel prices, agreed with Department of Energy. The procedures are not necessarily applicable directly outside the public sector, because they do not distinguish between capital and revenue accounts, nor do they need to include allowances for taxation. But there is no doubt that similar procedures, easily applied by staff throughout any organisation, help to create a situation in which individual initiatives in support of soundly based energy conservation measures, can readily flourish.

Improving Performance

In spite of the helpful factors described, work on energy conservation has not been altogether plain sailing. In particular, problems have been encountered in fuel monitoring, in tariff selection and control, and in operative training. Each of these is considered in turn.

FUEL MONITORING

Under the planned maintenance system, fuel consumptions are monitored and a check is kept on combustion efficiencies. But these procedures do not ensure that fuel is used economically in installations outside boiler houses and further checks are necessary to establish likely areas for economy.

Various methods are theoretically possible to meet this requirement, including the comparison of consumptions over a period, making due allowances for changes in weather and usage. But comparisons of this kind are never absolute and fall down because they fail to identify consistent misuse. An alternative approach would be to compare actual consumptions with a norm or target, but setting norms is difficult if not impossible in practice, due to the very wide energy variation experienced between the best and the worst cases of apparently comparable buildings serving the same kinds of function. A survey of a group of Crown buildings has shown that this variation can be as much as 40%.

The approach adopted has therefore been to produce a graph as shown in Fig. 1, in which fuel consumptions for each week of a heating season are plotted against the average outside air temperature for the week. If the inside air temperature is maintained at a constant level for the occupancy period, the relationship shown on the graph will be approximately linear, although the actual measured values will produce a considerable scatter of points due to the random effects of wind, rain, solar and internal gains, etc. However, the best fit straight line through the points plotted can be simply calculated using one of the available computer programmes and with the issue of meteorological data for average weekly air temperatures centrally, the whole operation can be carried out by clerical staff very easily.

When the graph has been completed, the line AB shown in Fig. 1 represents the fuel consumption when heat is provided for both space heating and other purposes not related closely to outside air temperature, such as domestic hot water services. The line BC represents the fuel consumption during the period when heat is provided for the latter only and the intersection of the two lines at point B should occur when the average outside air temperature corresponds to the 'no-load' condition. If point B occurs at an average outside temperature much greater than about 18–20 °C, then the building is normally overheating. In extreme cases of very poor performance, the plotted performance line would approach the horizontal.

This procedure is quite new and has yet to prove itself completely. But early indications are that in spite of the approximate nature of the results obtained, it provides a cheap and simple means of identifying

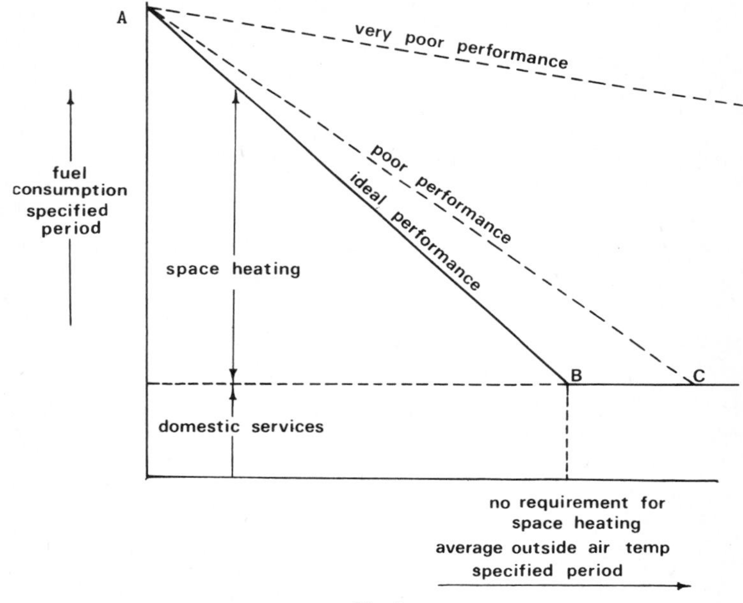

Fig. 1.

the worst cases, most deserving of attention. It is therefore useful in setting priorities and in concentrating attention on those areas where corrective action will pay the best dividends.

TARIFF SELECTION AND CONTROL

It is perhaps obvious to point out that tariffs need to be selected and used with care. But in a large organisation, with many thousands of separate agreements, centralised accounting procedures and subject to changes in organisation, staffing and responsibility, particular cases can easily be overlooked and considerable waste incurred.

A good example of the kind of problem that can arise, came to light during a survey into the running costs of air conditioned buildings, when it was found in the case of one large office building, that the refrigeration plant—not intended for winter use—had in fact been run up and operated during the winter period, following maintenance. In the electricity agreement for this building, the highest winter maximum demand in the November to February period, sets the level for the electricity charges throughout the following year and the effect of this one trial was therefore to double the average cost per unit of consumption and to incur an avoidable extra charge of almost £4000 at 1972 prices. The lesson here speaks for itself and in such cases, it is usually worthwhile to post large notices in plant rooms, as reminders to staff of the terms of agreements.

Further investigations also showed that in many cases, tariffs had not been amended to take the best advantage of changing circumstances, nor generally speaking, had the opportunity afforded by power factor correction been adequately considered. The position is not made easier by the different methods of charging adopted by the various Electricity Boards, where for example six Boards base their maximum demand charges on a kVA basis, while the remaining eight use a kW basis. In the latter category, there is still further inconsistency, because while an additional charge is often levied at the rate of 1% for each 0·01 change in power factor below a datum, the datum itself varies widely. Two Boards set the level at 0·85, four at 0·9, one at 0·93 and one at 0·95.

Fortunately, part of the answer to these sort of difficulties lies in the fact that accounts have been centralised and full details are held on one computer. It has therefore been possible to prepare new computer programmes for analysis purposes and as a first stage, an investigation was carried out to determine whether electricity accounts charged on block or flat rate agreements, could be improved by changing to maximum demand tariffs. In order to estimate maximum demand in those cases where consumption only had been recorded, a survey was carried out of 200 accounts already being charged on a maximum demand basis and having varying consumption patterns. Using a computer curve fitting programme, the results of the survey were used to deduce a relationship between electricity consumption and maximum demand, which in practice has proved to be about 80% accurate. Details of all cases found worthy of further investigation were printed out and about £300 000 of cost savings were achieved in

1973/74 following renegotiation. Table 1 shows typical results obtained in two large office buildings.

A more extensive review is now about to be launched as a second stage and will include the progressive recording of consumptions and costs incurred on all electricity and gas accounts. Calculations to determine the energy costs per unit of electricity or therm of gas will be carried out and the lowest cost tariff available will be indicated if not already adopted. Further calculations will also be carried out to

TABLE 1

	Building 1	*Building* 2
Annual charges on Maximum Demand tariff	£15 684	£14 388
Average cost per unit	1·205 p	1·319 p
Annual charges against previous block rate tariff for consumption	£19 630	£16 357
Average cost per unit	1·508 p	1·498 p
Annual saving	£3 946	£1 969
Reduction	20%	12%

indicate costs per unit floor area of the property under consideration, for comparison purposes in identifying the energy costs by building types.

Meanwhile, a separate survey of electrical accounts for civil Government buildings has shown that at least 100 properties had a power factor which made the provision of correction equipment cost effective, the capital outlay for the most part being recoverable within three years. The potential in the Defence estate is thought to be even greater. To cover the ground completely, a national contract has now been let on a survey, report, installation and commission basis, to cover approximately 1000 properties in the civil and Defence estates. Total expenditure on the contract is expected to be of the order of £250 000, with potential savings of about £100 000 per annum. In a typical case already reviewed under this contract, an overall power factor of 0·86 was being achieved at a maximum demand of 2260 kVA. An estimated 600 kVAr of capacitance is required to improve the power factor to 0·96, at which level the estimated savings in penalty charges amount to £3420 per annum, against a capital cost of installation of £6300.

OPERATIVE TRAINING

Only the very largest boiler plants are permanently manned by a full range of personnel with the right overall mix of skills necessary to meet all normal operational requirements. The majority of other plants are designed to operate automatically and are usually visited in rotation by semi-skilled personnel, who rely on specialist maintenance staff for all but the most straightforward running adjustments. In these circumstances, plants not infrequently operate at less than peak efficiency and recent sample checks have shown that fuel savings of about 5% can be obtained in many cases simply by giving closer and skilled attention to running adjustments. The use of skilled craftsmen rather than semi-skilled plant attendants in this situation, implies a cost penalty which is not however borne out in practice. In addition to the fuel savings made possible, a properly skilled man can undertake all first line maintenance tasks as part of his normal duties and is therefore completely cost effective, because much time consuming maintenance work by separate specialists can be eliminated. But a problem remains, because few skilled men have been trained with the necessary breadth of experience in combustion, boiler and heating controls, as well as in the more general electrical and mechanical maintenance fields.

To meet this requirement, new training courses are urgently needed. The hope must be that eventually they will be provided on a national basis by bodies such as the City and Guilds Institute. Meanwhile, the PSA, to suit its own needs, has had to devise and introduce a new series of intensive training courses to meet this requirement. The throughput of trainees is however limited by the size and number of courses that can be run in any period, and it will be several years before sufficient craftsmen trained to the new requirements become widely available in the organisation.

Energy Studies: Heating Installations

Space heating accounts for the most significant use of energy in Government buildings. The majority of existing buildings are naturally ventilated and heating is provided by conventional hot water systems served from central boiler plants. Oil still predominates as the fuel source, although many coal and gas fired plants are also in service.

From this starting point, the scope for major alterations to existing systems to improve energy performance in cost effective terms is limited. But none the less, considerable improvement is possible through straightforward action on draughtproofing and insulation levels; by closer attention to the needs of users, and if need be, their re-education; by improvements to control systems; and by better operational procedures and techniques.

As already stated, work in the control systems field began with the installation of optimum start controls to intermittently occupied office buildings in the civil estate. Approximately 1500 buildings have been converted and the annual savings resulting from this work now amount to more that £1 million, for about the same capital cost. This work is now being extended into the Defence estate, where because the incidence of intermittently occupied buildings is not so high as it is in the civil estate, a broader approach is being adopted. This will include a review of capacity control systems as well as the scope for optimum start controls. National contracts will be let shortly to cover the whole of the Defence estate in the United Kingdom and the programme will take about three years to complete. In this case, the potential savings are estimated at £10 million per annum for a capital outlay of the order of £20 million.

Improvements in capacity control may well involve internal temperature sensing. Many Government buildings heated by hot water systems are fitted with some form of compensator control which varies the water flow temperature to the heat emitters, according to changes in outside air temperature. However in practice, such air compensated systems usually have to be set up to satisfy the coldest area on the north face of the building and not infrequently, the remainder of the building is overheated and sometimes severely, in rooms subject to solar heat gain.

Attempts have been made to allow for the effects of solar gains by introducing zones controlled by solar and wind compensated outside detectors. But these rarely work satisfactorily and in most cases are difficult if not impossible to set up correctly, due to the varied effects of shading and wind. Indoor temperature sensing avoids these difficulties, though in many existing installations it raises other problems such as the need to modify pipework and heating circuits. A gain in fuel performance will not always pay for such modifications and it is recognised that installations employing continuous perimeter natural convectors pose a particularly difficult problem. Even so, considerable scope for improvement in this area remains and many authorities now agree that savings of about 15% are available from this source and our own studies bear this out.

In addition to improving boiler plant performance through closer attention by skilled operators, revised control methods for boiler plant particularly to deal with medium and light load conditions, are thought to offer further scope for worthwhile savings. Studies and tests are therefore being conducted in an attempt to improve load sensing, by controlling plant from the heating system return water temperature, rather than from the boiler flow water temperature, as one possible means of eliminating unnecessary cycling of on/off burners, under medium and light load conditions. Other aspects of standard practice are similarly being re-examined, including the maintenance of constant boiler flow water temperature at all loads in systems provided with mixing circuits, the sequencing control of multi-boiler installation by reference to boiler flow water temperature and the use of variable water volumes through many systems. It cannot yet be stated with certainty where these enquiries will lead, but early indications are that at least a further 5% fuel saving is achievable in this general area.

Further Work in the Electrical Field

A survey of electricity consumption patterns in a sample of sixteen medium sized non-air conditioned Government office buildings has shown that approximately 65% of the electrical energy consumed is used in lighting, with the balance being taken by lift installations and other general power services.

Figure 2 shows the breakdown of electricity consumption over a 24 h period, for one office building in the London area. In this case, since 70% of the electrical energy was consumed in lighting, a special study of this aspect was mounted. It showed that lights were in use for a typical period of about 3500 h per annum, compared with a theoretical need in normal shallow plan naturally lit buildings of not much more than 1000 h. The difference in these figures indicates the considerable scope for energy economy through voluntary action by building occupants, in the economical use of lighting.

It would, however, be wrong to lay the whole blame for uneconomic operation on the building occupants. Further studies have shown that a major factor in the cost of lighting of office buildings in particular, is

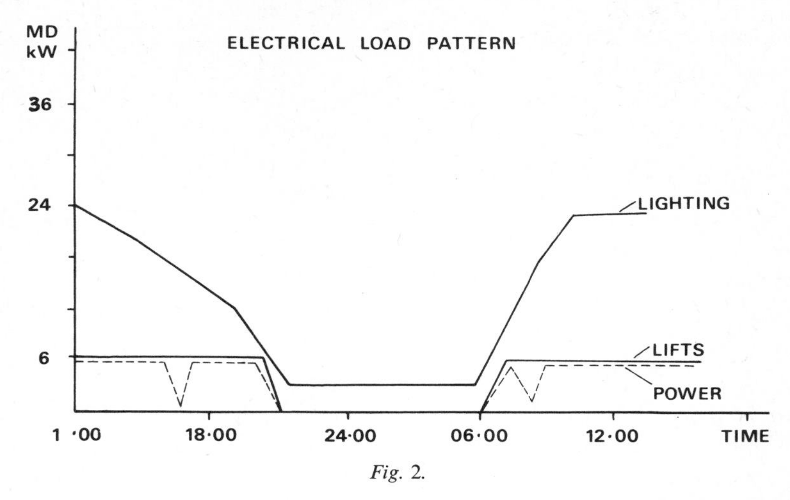

Fig. 2.

the practice of cleaning during the hours of darkness. It is commonly found that well disciplined staff will switch off all lights when they leave their desks at the end of the day, but an hour or two later when the cleaners move in, the complete lighting system tends to be switched on again in order to facilitate the flow of cleaning work. The effects of this practice are clearly shown in Fig. 3, which plots the electricity consumption of a typical office building over a 24 h period. The dotted lines in Fig. 3 have been inserted to show the probable effect on

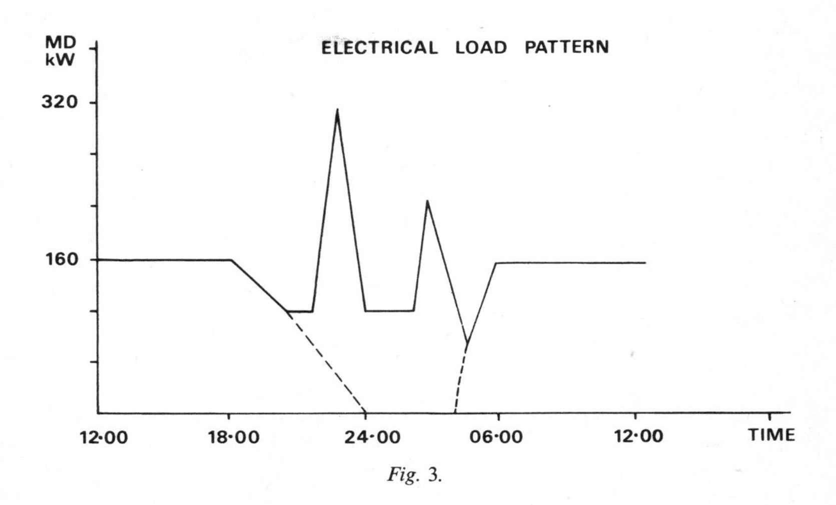

Fig. 3.

consumption demand, if cleaning were to be restricted to the beginning and/or end of the occupancy period only. The likely effects of such action on both electricity demand and costs are further illustrated in Tables 2 and 3, Table 2 showing the recorded position in a particular building during the period September 73 to August 74,

TABLE 2
OFFICE CLEANED OVERNIGHT

Annual unit consumption	2 059 440 kWh
Highest MD recorded	800 kW
Annual Unit Cost	£16 908
Annual MD charges	£6 772
Total electricity cost	£23 680

and Table 3 showing the estimated savings potential, for the same building. It is notable that this single action can save 18 % on energy, 23 % on maximum demand and result in a total net cost saving of 19 %. From this experience, the lesson to be learned is clear cut and at the very least, future cleaning contracts need to be arranged with the energy cost implications taken into account.

TABLE 3
OFFICE CLEANED AT BEGINNING AND END OF A WORKING DAY

Annual unit consumption	1 688 741 kWh
Highest MD recorded	660 kW
Annual Unit Cost	£13 820
Annual MD charges	£5 200
Total electricity cost	£19 020

Further improvement in lighting performance is often possible by rearranging circuits and switching. In many existing buildings, the present switching arrangements are often unable to provide economic control, because they do not discriminate between the different needs of supplementary daytime lighting and full artificial lighting after dark. Various alternative switching configurations are possible to provide increased discrimination and there is little doubt that in future the most efficient installations will include provision for:

changing over lighting levels from day to night operation;
reducing supplementary lighting near windows on bright days;
providing nominal lighting levels for cleaning purposes.

Control of lighting installations engineered to meet these new requirements can be arranged either manually or automatically. Automatic control by means of change-over switches and photo cell sensors, implies a degree of loss of amenity, and can bring adverse reactions from staff. However to some extent these can be overcome by providing individual local desk lighting and such provision will in any case often lead to further energy saving by individuals, many of whom prefer to work under local rather than general lighting conditions. Even so, care still needs to be exercised in the adoption of automatic control systems both because of their effect on maintenance costs and because severe disruption of work can be incurred if the controls malfunction or operate oversensitively to produce too frequent switching or dimming. Further investigations are therefore in progress on this and related problems.

In another area, tests have shown that there is not much scope for energy economy in lift installations. Most modern lifts are already engineered to regenerate when a full or nearly full car descends or when an empty or nearly empty car ascends, pulled up by its counterweight. In addition, if a lift installation has been properly designed and sized for the correct building population in the first place, there is seldom any point in attempting to achieve economy by taking one or more of a group of lifts out of service. Generally speaking all that happens is that the remaining lifts run more frequently, and power consumption is hardly affected, though waiting time rises significantly. Even if such action were to produce some modest saving in energy, it is arguable that the man hours lost through the reduced lift service are worth considerably more than the insignificant energy saving achieved. Overall, it is not a course to be recommended.

Catering Installations

Bulk catering installations are provided in many Government buildings, both as part of the various messes serving the Armed Forces and in support of the civilian staff restaurants operated by the Civil Service Catering Organisation. Since catering installations are by their nature energy intensive, a number of investigations are being undertaken into methods of reducing energy consumption in this field. The potential level of savings thought to be achievable throughout the civil and Defence estates is estimated at about £3 million per annum.

Several areas are under investigation, of which the most important is possibly the development of an automatic sensor to detect whether boiler rings are covered by pans. Considerable energy is wasted by catering staff who tend to switch on boiling rings at the beginning of a working shift and to leave them on throughout the period, whether cooking is in progress or not. A development contract has therefore been let to study gas systems in detail as a first phase, and a prototype sensor has now been submitted to British Gas for evaluation and approval.

On the electrical side, trials have recently been completed of a new bulk storage water boiler for tea making in offices, which has been designed to include a thermostatically controlled main element, supplemented by a boost element which the user controls by means of a push button. In use, the temperature of the water is maintained just below boiling point and it is only brought to the boil by means of the supplementary element as boiling water is required. The trials of this unit have shown that it uses 60% less energy than the normal continuously boiling tea urn and, as a further bonus, it produces less scale, is less subject to maintenance problems and it improves the flavour of the tea.

Further work being undertaken includes the development of new automatic ignition systems for gas appliances to eliminate pilot lights; the provision of manual reset timers for cooking grills, to allow cooking times to be predetermined and controlled automatically; together with tests on hot cupboards at serving counters, where it has been found that cabinets with very high insulation levels can show an energy saving of 18%, compared with the normal commercial equivalent.

The catering field is a fruitful one for energy conservation, in another sense too, when it is borne in mind that however efficient or inefficient particular items of equipment may be, it is often possible to recover the 'wild' heat from kitchens and re-use it by means of thermal wheels or similar techniques, to warm the adjacent dining and bar areas.

The Design of New Buildings

The rate of replacement of existing buildings is not so rapid that new buildings and facilities can make a significant contribution to energy conservation in the short term. However, with an average building life expectancy in excess of sixty years, no effective energy conservation policy can afford to ignore the longer term prospects and standards being set by new designs.

In the new design field particularly, energy conservation is not the prerogative of engineers of any kind, or of architects, or of surveyors, and it will only reach its full potential if all the disciplines work closely and effectively together. It is arguably the final spur needed to achieve full multi-disciplinary working and in this context perhaps the greatest contribution by the PSA will prove to be the outcome of joint discussions, now underway with the Association of Consulting Engineers, aimed at restructuring fee scales. One of the most important elements in this exercise, covers the terms and conditions needed to provide for full inter-disciplinary consultation at the conceptual stage of design, when vital decisions affecting the subsequent energy performance of the building are taken.

On-going work in new design reflects generally the same theme and multi-disciplinary collaboration is being actively encouraged, to achieve the best overall energy performance in individual cases. Design branches have been asked to report results, to determine whether energy targets can be established, in order to limit maximum and annual energy demands. As this data begins to flow, further consideration will be given to the practicability of basing targets on floor area, building volume or number of occupants. Considerable controversy exists about the merits of this procedure, compared with the alternative of setting elemental standards of thermal construction for walls, roofs and floors and for the control of air infiltration. The use of air conditioning in some buildings, raises the further question of whether the targets in such cases should be different from those for naturally ventilated and conventionally heated buildings, or whether heat recovery and other techniques should be employed to keep within the same overall target. It is too soon yet to pronounce on the outcome of this activity, but the possibilities will continue to be studied and other factors will be brought into account as necessary. These include the allowances to be made for intrinsic heat gains where, in the case of factories or other process buildings, very different considerations from those applying to offices and living accommodation can arise.

Multi-disciplinary action is also being encouraged in the computer field. At present most disciplines have their own programmes and these are very useful for providing quick and accurate analyses, whether for cost estimation, structural performance or thermal evaluation purposes. They avoid much tedious and time consuming manual calculation and enable a thorough examination of alternative solutions, even on urgent projects, where in the past time has often precluded the necessary studies in depth, to the detriment of the subsequent performance of the building, particularly in energy terms. But the nature of these programmes means that they have generally been applied and used separately by the professions on a piecemeal basis and a new approach is therefore being developed aimed at making the computer the centre piece of a project team meeting, where members of the team, including the client, would be able to obtain initial and operating costs, together with energy demands and other main parameters, for various changes in building geometry. Operated in this mode, the computer would be fed with multi-disciplinary base data, related to a particular design or proposal in which relevant materials, dimensions and conditions would be described. A visual unit would display the building configuration, with individual application programmes called up to give on the spot results to the team. By changing the base data at will, the team would be able to produce sets of results from which an optimised outline plan could be developed. Work on this system is now proceeding, and since it draws heavily on programmes already fully developed, including those for displaying building configurations, it is hoped that worthwhile results can be produced for the minimum additional investment.

Conclusions

In a paper of limited length on a subject as broad as energy conservation, it is only possible to include a 'snapshot' of samples of activity to show the complexion of some of the problems encountered, of successes, failures and areas where further work is needed. It is hoped that the sample chosen is not too random and that sufficient data have been included at least to raise questions and set us thinking. In most cases, there is nothing peculiar about Government buildings, except possibly their quantity, and the majority of points covered should generally find an echo in other organisations, public or private, equally concerned to obtain the best value for money and to make an honest and worthwhile contribution towards conserving the dwindling world resources of fossil fuels.

The Appendix has been produced as an aide-memoire and as a summary of the major areas of short term action discussed. But a note of caution is necessary, in case anyone should assume that the percentage savings shown in the Appendix are additive. In most cases they are not, because the various measures listed inter-react one with another. They do indicate however that the overall savings target of 30 % set by the PSA is unlikely to be an impossible dream and indeed it is noteworthy that in the first full year of the new programme, over £12 million of fuel saving has been recorded. A good portion of this has been due to the exceptionally mild winter last year, but with due allowance for weather, some £8·5 million of saving can fairly be ascribed to the energy conservation measures already implemented. This is equivalent to 10 % of the energy used across the board before the present programme was launched.

In conclusion, it should be remembered that there is no philosopher's stone in the search for improved energy economy. Only constant attention to detail in all areas and improved co-ordination by all professions in the industry will bring its reward. Even then, the long term results will continue to depend on the way in which building users respond and on the quality and effectiveness of the maintenance and operation organisation provided.

Appendix—Civil and MOD Estate—Existing Buildings

Type of conservation/ financial benefit	Potential saving in % of sectorial consumption (years)	Costs/ Benefits	Type of policy and action required
(a) Running boiler plants at optimum efficiency	5% of fuel	small cost compared to savings	information, training, encouragement, monitoring performance, accountability
(b) Less non-productive idling of boiler plant	5% of fuel	small cost compared to savings	information, training, encouragement, monitoring performance, accountability
(c) Introduction of Optimum Start Control to MOD Estate	10% of fuel	return on investment in 1–2 yrs	identifying suitable buildings for conversion, conversion contract, monitoring performance
(d) Adjustment to existing heating controls—MOD and civil Estate	10% of fuel	small cost compared to savings	monitoring performance to highlight large wastage. Corrections to controls. Training staff in control technology. Accountability.

(e) Installation of heating controls to inadequate system—MOD and civil Estate	10% of fuel	return on investment in less than 1 year in most cases	monitoring performance to highlight large wastage. Installations of controls. Training of staff in control technology. Accountability.
(f) Improvements to electrical power factor	2% of electricity cost	return on investment in less than 3 yrs	Identifying poor power factor. Fitting power factor correction devices. Monitoring performance. Accountability.
(g) Reduce excess lighting by staff action	20% of electricity	small cost compared to savings	Encouragement and publicity. Monitoring performance.
or			
Reduce excess lighting by technical improvements in switching and controls	20% of electricity	return on investment in 1–4 yrs	Development of new design and control concepts. Implementation and monitoring performance.
(h) Improve thermal insulation and temperature control	20–25% of fuel	return on investment in 5–10 yrs	Revised insulation and control standards. Implementation of schemes giving greatest cost benefits.
(i) Automatic controls on catering equipment	20–25% savings on modified equipment	return on investment under 4 yrs	Development of fuel savings devices and implementation of the scheme after Gas Board approvals.

REFERENCE

1. *Planned Maintenance and Operation of Mechanical and Electrical Services.* Engineering Instruction M8: HMSO.

Discussion

R. G. Borthwick (*UKAEA*): I would like to talk not about Atomic Energy but savings in more mundane fuels! May I commend the use of the straight line relationship between outside temperature and fuel usage indicated by Mr. Johnston and illustrated in Fig. 1. I have found this a very useful tool over the last 10 years, not only as a measure of fuel conservancy but having several advantages in assessing the heating of different buildings. When the points are first plotted, one finds a scatter but reasons for the scatter can be found in most cases. East and North winds cause points to fall below the line indicating increased fuel usage whereas solar gain causes points to fall above the line. After some experience allowances may be made for such variables.

Each type of building produces a line with a different slope, a low slope indicating less variation in fuel usage for heavy, well insulated buildings. The graph for this type of building intersects the temperature axis at a lower temperature than a lightweight building *e.g.* intersection at 18 °C for a light building and as low as 15 °C for a well-insulated building. This reading indicating that for the latter type of building a 20 °C inside temperature may be maintained without heating at lower outside temperatures.

During the 'three day week' fuel crisis of winter 1973/74 we found this performance line very useful to demonstrate the effectiveness of our fuel economy measures, some of which were only temporary but others became permanent. Since then other fuel economy measures, such as improved insulation, improved boiler operation and a reduction of inside temperature from 20° to 18 °C, have shown the effective saving as the position and slope of the line changed. It was always difficult to explain that despite savings made in fuel consumption, we used more fuel in a cold winter than in a mild one and savings were not obvious. The straight lines for each heating complex plotted over the years show these savings in unmistakeable terms.

H. A. Rudgard (*Shell International Petroleum Company Limited*): Energy has been saved for the past six years in Shell Centre and other buildings within the Central London Maintenance Association. The 30% saving potential (quoted by Mr. Johnston) has now been achieved. In the last two years at Shell Centre 15% of electrical energy costs and 18% of fuel oil costs have been saved. Savings are now getting past the house-keeping stage and supervisors are looking towards monitoring conditions in the building and switching plant according to the conditions. Why, for example, operate an air conditioning plant continuously in an area in which the occupation can vary from the extremes of being empty and fully populated.

S. Mulcahy (*Steenson Varming Mulcahy & Partners*): It is emerging that the outcome of the conference could very valuably be the identification of three targets—the short-term, the medium and the long-term targets. I think more than anything what has to be identified is who is to do what, where and when? Into this we have to bring the training ideas put forward by Mr. Johnston and the Institutions who have arranged this conference have very heavy responsibilities here. What is needed I think is a 'retrofit' of those of us who are already in the profession and in the operations. It has been said that we are stuck with the existing buildings—I think we are also stuck with the existing designers and operators.

E. G. Brooks (*Institution of Heating and Ventilating Engineers*): I wonder if the message of urgency has really come over. Mr. Mulcahy came nearest to this sense of urgency together with Mr. Nelson's earlier comment about 'survival'. We have an extremely important task to deal with, and we have got to convince the Government of the need to save. We talk quite freely about 'what is the Government doing?', should not we be talking about 'what we are doing to convince the Government' for that is the way it has got to be! The Government already has an Advisory Council on Energy Conservation chaired by Professor Sir William Hawthorne. I hope we are going to achieve something which will convince the Council and the Government to save energy.

Dr. J. Gibson (*National Coal Board. President–Institute of Fuel*): As far as I am aware the Hawthorne Committee does not contain a heating and ventilation engineer or a fuel technologist, which are serious omissions. The Institute of Fuel is, of course, a qualifying body, but our resources are too small to allow us to carry out work which is the function of the Department of Education and Science. It is their job to provide courses of instruction; it is ours to prescribe the content of these courses. The cessation of City and Guilds courses has been mentioned with regret, which we share. Unfortunately, education is governed by bureaucrats and, as I see it, they seek the least cost line irrespective of national need. The Institute has submitted evidence and appeared before the Select Committee on Science and Technology on Energy Conservation. We offered to help with training and education and I wrote a letter to 'The Times' saying that we should do something now, before it is too late, to educate the technologists, engineers and scientists we need to carry out an effective energy programme. Nothing has happened. I have now written to the Secretaries of State for Energy and for Education and Science about the need for education, offering help, but no one seems to want to move. I am glad, therefore, to hear Mr. Johnston say that we need more education for energy conservation.

G. Nelson (*Oxford Polytechnic*): I support Mr. Brooks when he suggests that we should, as responsible professionals, press the Government into meaningful action. Although I must express my doubts about the value of this action if their most recent proposal suggesting the formation of a 'flying squad of trouble shooters' who would investigate the waste of energy is a criterion. Such a group existed—The National Industrial Fuel Efficiency Service—and was disbanded quite recently as a result of Government policy.

D. N. Soane (*Surrey County Council*): I would like to ask the question, 'is the Government likely to increase the cost limits allowed for buildings in the public sector, to allow for energy conservation measures?' In my view it would appear unlikely that any immediate increases will be recommended. Much more awareness of the situation will be needed, and much more publicity to get the message through to some Government Departments. For example, at one time, and for some time, although not now, the Department of Education and Science allowed the same cost limit for a Technical College as an ordinary school. I would call for realism in setting cost limits particularly where energy conservation is concerned.

W. J. Ablett (*Ministry of Defence*): I would like to confirm Mr. Johnston's suggestion on the need for regular tariff review. The Ministry of Defence have been saving, on average, £100 000 per annum in Germany, simply by reviewing and changing tariffs.

One of the major points emerging from this conference requiring clarification is this—are we spending money primarily—economically to save energy or are we spending money to reduce revenue cost with energy being a secondary consideration. Three different organisations have quoted 'pay-back' periods on their capital expenditure, *i.e.* 2, 7, and 10 years. I wonder what energy conservation results would have been achieved if the incentive of the financial 'pay-back' situation did not exist.

Over the years, I have discovered that heating and air-conditioning control systems in buildings generally leave a lot to be desired. Most control systems can be classified under three headings:

(a) the wrong system was chosen in the first place to meet the users' requirements.
(b) that the control system installed was correct but is not functioning correctly due to lack of regular checking of the system and
(c) a system has been installed but was never properly commissioned. It has never worked and never will work.

Now on my attempt to examine priorities of expenditure showing the best rate of return in energy conservation, I find that improved control systems comes out near the top. I wonder if anyone has carried out similar investigations and could provide an indication of their results.

Two other energy saving points. The first is that without spending money, by simply arranging to turn off a simple control known as the radiator wheel-valve during the summer months, and subsequently leaving it for the room occupant to turn it on produces 3 results:

(a) the twice annually operation of the valve prevents it from becoming corroded.

(b) the heating in each room is likely to be turned on at different commencing periods of the heating season, reducing demand and
(c) the room occupant becomes conversant in the use of the valve instead of opening the window.

The second is to indicate the possibilities of aerial night photography using infra-red film. The film shows heat losses from buildings and underground heating mains. It is staggering to see the enormous amount of heat that is being wasted.

Dr. J. Gibson: Government has not paid sufficient attention to the need for education and training; it has not taken the opportunity to get people of different disciplines together to optimise a situation; it has set goals that are too low. These conclusions have come out of the discussion and will continue to stimulate us. I think we have established that once the obvious energy that costs nothing has been saved, it requires money to carry out further conservation. We must invest, but investment will only be worthwhile if it is carefully planned and properly implemented.

D. N. Soane (*wrote*): Many stimulating ideas emerged from the conference, but I would have liked to see more emphasis on measures that can be taken immediately, without undue cost, to ensure our own 'houses' are in order.
We have for many years operated measures that are considered essential to economy such as regular fuel returns, acting on any excessive consumption by ensuring Plant Reports cover such items as time clock and thermostat settings, condition of boiler flues, frequency of cleaning, windows open, complaints regarding over or under heating etc; but much more can be done. We must try to ensure new installations, and what has been inherited from the past, are operating properly. All too often it is found not to be the case. We all know of controls that have not been properly installed and do not work as intended. Where we have had pilot 'optimum start' controllers installed, the savings effected have not so far been of the order suggested at the conference; as all buildings, other than residential homes, shut down at night. The majority which are on external compensated control, have an outside frost thermostat bringing in the heating pumps should the outside conditions fall to 1·7 °C and a return thermostat set at 7·2 °C to bring the heat source into operation. The clock switching is set to reasonably close limits and there is an internal boost thermostat which switches on the outside compensator, when the internal design temperature is reached. Much surely depends on how the buildings have been operated in the past and the indiscriminate application of optimum start control does not immediately appear justified. However, there does appear to be some benefit to be derived from installing them in lightly constructed buildings with low thermal capacities.

H. Edge (*Lancashire County Council*) *wrote:* It has been stated that new buildings constitute only 3% of the present building stock. Presuming that Energy conservation is an urgent and immediate necessity, it is apparent that the responsibility for its implementation lies more with building and plant engineers, preferably Energy Conservation Engineers with the necessary experience and expertise, rather than Architects, Design Engineers or Accountants, who generally are concerned with the design of new buildings. I suggest that with the apparent lack of demand, through advertisements in the press, for additional Energy Conservation Engineers, with the possible exception of Local Authorities, more organisations should be examining the possibilities of employing such staff to operate in the field, rather than in a design department.

The following brief information on the results of employing such staff in Lancashire County Architects Department may be of interest. Resulting from earlier research into building energy costs, one Fuel Efficiency Engineer was appointed in 1967, quickly followed by the establishing of a Fuel Efficiency Unit in 1969, consisting of ten engineers, recruited from various energy industries and consulting organisations. During the four year period 1969 to 1973, with approximately 2000 schools and other premises the total energy bill was approximately £2 500 000, rising to £5 000 000. A total saving exceeding £1 000 000 was obtained with salary costs less than £100 000 during these four years. The savings averaged 14% with negligible capital costs on minor items.

Since Local Government Reorganisation, the Authority has been smaller in size and six Fuel Efficiency Engineers control an energy account of approximately £5 000 000 per annum. These engineers spend the majority of their time in the field and each of the Authority's buildings has been surveyed at least once. During a survey:

(1) The building energy costs are made available to the building manager or headmaster, and utilisation aspects discussed.
(2) Electricity and Gas tariffs are corrected where necessary and costs have been halved as a result, in extreme instances.
(3) Boiler plant and other engineering services are tested and checked for correct operation, adjustments made, controls sealed and maintenance or service arrangements put in hand, if necessary.
(4) Building Staff are advised and instructed on correct operation of the plant.
(5) Suitable publicity material, such as posters, light switch stickers, etc. are provided.

There has been reference to the installation of optimum start equipment to heating plant, at a cost of approximately £500 to £700 per building. It is perhaps interesting to note that Lancashire County's Fuel Efficiency Unit Staff have themselves, during the past eight weeks replaced 360 time clocks with improved modified types costing £7000 at 360 schools. The point to be made is that with current financial restrictions a choice is often available, such as whether to install 10 Optimum Start Installations or 360 new time clocks at a corresponding number of premises. A reduction of oil consumption of up to 13% is anticipated at these premises. To illustrate again the potential energy savings with minimum expenditure, an exercise at six schools, carried out earlier this year during a three week period, revealed that the application of

'Good Housekeeping Techniques' by school staff supplemented by some technical control, reduced energy consumption by (1) lighting–64%, (2) power–26%, and (3) fuel oil–23%. I suggest therefore that effective Energy Conservation requires only the employment of suitable Fuel Technologists or Energy Conservation Engineers, who have adequate experience and flair in these techniques, with a negligible capital expenditure.

7

Energy Conservation Studies for New and Existing Buildings Produce Guidelines and Standards—Tools for Decision Making

FRED S. DUBIN and JOHN BARNABY

Synopsis

The demand for energy in the United States is rapidly increasing while the domestic fuel supply of fossil fuels is rapidly decreasing; new ways for reducing the gap between supply and demand are being sought. Energy conservation in buildings is one means of reducing the demand for energy. Therefore, in order to promote energy conservation in buildings, to establish guidelines for energy conservation programmes and standards for building codes, and to encourage legislation which determines and limits the consumption of non-renewable resources, energy conservation studies have been commissioned and funded by United States government agencies as well as groups in the private sector; these studies have been conducted primarily by professional engineers and architects.

This paper describes three energy conservation studies for the government and one for a corporation, the elements of several subsequent projects involving energy conservation studies and/or designs performed specifically to conserve energy in buildings. The four studies are:

(1) Energy Conservation Demonstration Project, Federal Office Building, Manchester, New Hampshire: sponsored by the US General Services Administration of the United States government. The purpose of the project (which includes The Study in addition to the ultimate design, construction and use of the building) is, according to the General Services Administration, 'to (a) dramatise the firm

EDITOR'S NOTE. This paper was originally prepared in Imperial Units; wherever possible the units have been converted to Système International (SI) but the original units have been retained in parentheses.

commitment of PBS/GSA to the conservation of energy in the design, construction and operation of federal buildings; (b) provide a laboratory for the installation of both recognised and innovative energy conservation techniques/equipment; and (c) inspire others in the building construction industry to pursue energy conservation as a goal.'

(2) GSA Energy Conservation Guidelines for New Office Buildings: also sponsored by the General Services Administration. This study, completed in 1974, describes the relationships between energy consumption in buildings and the climate, building envelope, mechanical and electrical systems and building use. Energy conservation options are listed in order of priority by climate and system. The manual develops perfomance standards and establishes an energy budget goal of 625 MJ/m^2 yr (55 000 Btu/ft^2 yr) at the building boundary and 1249 MJ/m^2 yr (110 000 Btu/ft^2 yr) raw source energy for new GSA buildings, regardless of location in the country. The GSA issues these guidelines[1] to all the design professionals whom they engage for new projects.

(3) Energy Conservation Studies for Existing Buildings for the Federal Energy Administration: sponsored by United States Federal Energy Administration. The purpose of the project was to study energy usage in existing buildings, to research and develop energy conservation measures, and to prepare a set of manuals for use by the public and professionals, detailing the energy conservation options available. As a result of this study, two manuals were prepared and are available,[2] one (subtitled 'ECM-1') is a manual designed to help building owners, managers, operators and occupants to conserve and manage energy usage now, without the need of investing a significant amount of money, the other ('ECM-2') is a working tool for engineers and architects who are commissioned to analyse energy usage and recommend and design appropriate measures to conserve energy in existing buildings.

(4) Connecticut General Life Insurance Building, Bloomefield, Connecticut: An energy management programme commissioned for the company's existing home office building. The studies conducted over a period of time included analyses of the loads due to the building envelope and heating, ventilating and air-conditioning distribution systems, an energy audit of the fuel and electricity used for heating, ventilating, air-conditioning, lighting, illumination and power; and analyses of the design, condition, and operation of the mechanical

and electrical primary energy-conservation equipment. The quantitative results of subsequently retrofitting the systems and modifying their operation in accordance with designs performed are discussed.

The four energy conservation studies and designs mentioned above, were instrumental in providing an expanded body of knowledge for subsequent work by the firm.

The studies and/or designs of five of the energy conservation projects which have been completed within the past year are briefly described, and the energy conservation features of each are summarised. The projects include:

The Administration and Research Building for the Cary Arboretum of the New York Botanical Gardens in Millbrook, New York;
The Management Center of Argonne National Laboratories in Argonne, Illinois;
The Minnesota Mining & Manufacturing Corporation Research and Development Center in St. Paul, Minnesota;
The US Home Corporation's 'Resource Saving House' in Clearwater, Florida; and
A Study to Forecast Power Requirements and Reduction in Peak Demands and Consumption for Long Island, New York.

The alternative methods of decreasing the gap between domestic energy supplies and demand are enumerated in the 'Introduction', and the major advantages of energy conservation studies and the energy conservation practices which result are described.

Introduction

Energy demand in the United States has grown rapidly since the turn of the century, increasing at an annual rate of 4% to 5% during the last decade. As recently as 1950, the United States was completely energy self-sufficient; it readily met its burgeoning power needs with relatively cheap and abundant domestic fuels—first with coal, and then, increasingly, with oil and gas—as well as with hydro-electric power. But by 1960, just ten years later, imports of crude oil and

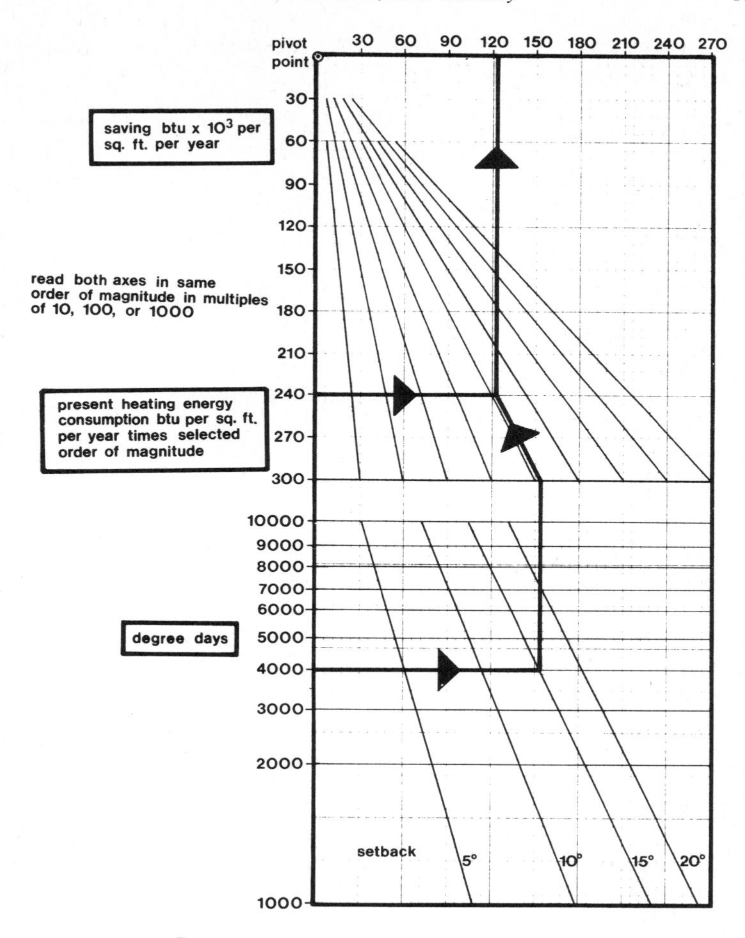

Fig. 1. Heating: energy saved by night setback.

petroleum products accounted for 15 % of total domestic petroleum consumption, and by 1973 imports had escalated to 35 % of the total.

And now, we in the United States find ourselves with less fuel of our own, with higher prices for fuels from abroad, and with an ever increasing need for more energy; a national cry goes up that we should once again become energy self-sufficient.

So that we may re-attain energy self-sufficiency, our government and industry are considering several options open to us for increasing our domestic fuel supply to meet our energy needs. Briefly summarised, these are the options:

(1) We can increase our production of fossil fuels, especially coal, on our own continent; but this solution assures early depletion of our fossil fuels and guarantees a more serious air and water pollution problem as a result of increased combustion.
(2) We can increase fossil fuels production through the exploitation of shale oil, coal gasification and liquefaction, and other synthetic processes; but this solution places an additional burden on our already strained water resources and will cost far more than present domestic fuel supplies.
(3) We can increase nuclear power production; but simultaneously, of course, we will increase the hazards due to radioactive waste disposal limitations and other potential dangers still inherent in nuclear power plants.
(4) We can undertake to develop programmes which increase energy from what we have come to call 'alternative sources', *i.e.* solar, wind, ocean thermal differences, and geothermal power. This solution, when thoroughly developed, may well prove to be one of the most reliable and non-polluting possibilities for supplying a large portion of the energy needed.

But another major option open to us, which is too often dismissed, is to lower our demand for energy, in an effort to assure ourselves and future generations of an adequate supply of useful fuels. To lower our demand for energy, many efforts can be undertaken which do not require unmitigated curtailment of energy use, but do require reducing waste and making wise choices for allocating resources to specific uses in a more efficient manner.

During the past few years we have been engaged in conducting research and preparing studies for public and private clients in an effort to reduce the amount of energy used in buildings and to encourage others to consider energy conservation in buildings as a viable alternative to increasing the supply of energy.

The value of reducing energy demand in buildings can be more fully appreciated when the following facts are considered: About 35% of all energy used in the United States is consumed directly in buildings; another 6% consumed off-site in facilities to support buildings for such purposes as sewage treatment, water supply, and solid waste management; approximately 7% more is used to process, produce

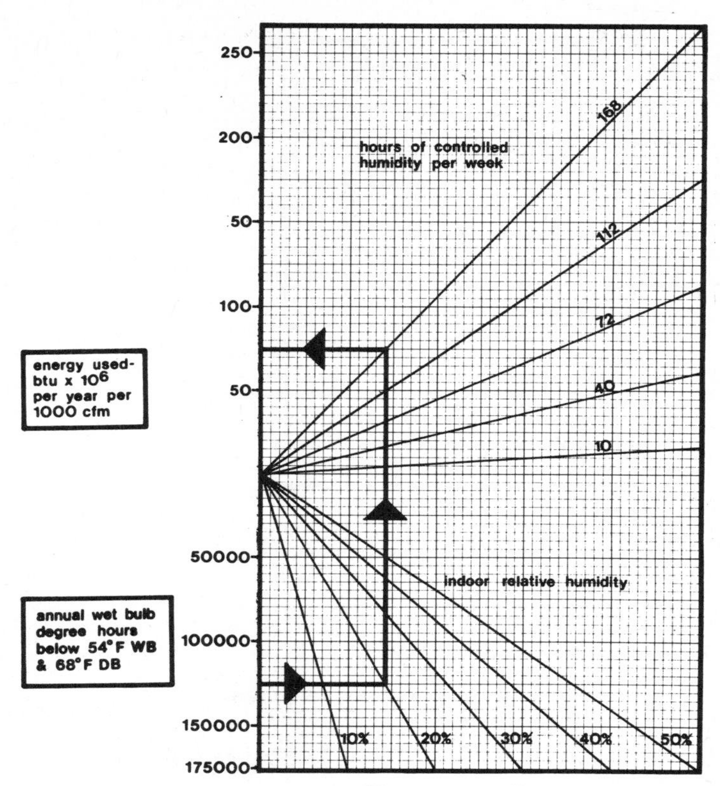

Fig. 2. Heating: yearly energy used per 1000 *cfm to maintain various humidity conditions.*

and transport materials used for building construction—all told, about 48 % of all energy used is in and for buildings. The major energy uses for the operation of the environmental control systems in buildings in the United States are: 57 % for space heating, ventilating and air conditioning; 33 % for the operation of electrical motor driven equipment, domestic water heating, home appliances and office equipment; 10 % for lighting.

If the energy used in buildings were reduced by just 25 %—and this would not be difficult to do—a saving of an equivalent of three million barrels of oil per day could be effected. Some existing office buildings have already achieved a 30 % reduction in energy use without requiring a significant investment. Analyses indicate that

TABLE 1a
YEARLY HEAT LOSS/SQUARE METRE THROUGH WALLS

City	*Latitude*	*Solar radiation MJ/m² day*	*Celsius base degree days*	*Heat loss through walls MJ/m² yr*											
				North				*East and West*				*South*			
				U = 2·21 *W/m² °C*		U = 0·57 *W/m² °C*		U = 2·21 *W/m² °C*		U = 0·57 *W/m² °C*		U = 2·21 *W/m² °C*		U = 0·57 *W/m² °C*	
				a = 0·3*	a = 0·8*	a = 0·3*	a = 0·8*	a = 0·3*	a = 0·8*	a = 0·3*	a = 0·8*	a = 0·3*	a = 0·8*	a = 0·3*	a = 0·8*
Minneapolis	45 °N	13·6	4657	845·1	802·3	232·3	219·6	801·3	706·7	220·1	190·6	750·2	582·5	205·6	153·6
Concord, N.H	43 °N	12·6	3889	780·8	736·2	214·6	201·2	734·4	628·7	201·5	170·0	678·6	495·9	185·9	128·8
Denver	40 °N	17·8	3491	651·1	612·6	178·9	168·3	610·1	510·3	167·6	138·5	553·9	387·2	152·2	99·0
Chicago	42 °N	14·7	3419	664·5	628·6	182·6	172·7	627·1	541·4	172·3	146·1	575·6	424·0	157·2	110·6
St. Louis	39 °N	15·7	2722	511·5	478·6	140·6	131·3	476·7	399·6	131·0	107·6	432·0	299·2	118·4	75·6
New York	41 °N	14·7	2706	521·3	487·7	143·3	134.0	486·5	401·6	133·7	108·9	435·9	286·5	119·8	72·7
San Francisco	38 °N	17·2	1675	264·1	239·8	72·6	65·9	237·5	177·5	65·3	46·8	192·5	111·4	52·7	19·8
Atlanta	34 °N	16·3	1657	305·7	281·7	84·0	76·9	277·9	218·1	76·3	57·9	234·4	140·8	63·2	29·4
Los Angeles	34 °N	19·7	1145	112·4	97·1	30·9	26·7	95·3	65·4	26·2	14·9	69·7	34·5	17·3	1·8
Phoenix	33 °N	21·8	981	134·7	119·6	37·0	32·7	116·8	83·1	32·1	20·6	91·7	52·5	23·4	6·3
Houston	30 °N	18·0	889	165·7	147·1	45·5	40·4	146·4	106·5	40·2	26·7	123·5	76·8	33·0	13·0
Miami	26 °N	18·9	78	2·4	1·2	0·08	0	1·05	0	0	0	0	0	0	0

* 'a' is absorptivity or absorption coefficient

TABLE 1b
YEARLY HEAT LOSS/SQUARE FOOT THROUGH WALLS

City	Latitude	Solar radiation Langley's	Degree days	Heat loss through walls Btu/ft² year											
				North				East and West				South			
				U = 0·39		U = 0·1		U = 0·39		U = 0·1		U = 0·39		U = 0·1	
				a = 0·3*	a = 0·8*	a = 0·3*	a = 0·8*	a = 0·3*	a = 0·8*	a = 0·3*	a = 0·8*	a = 0·3*	a = 0·8*	a = 0·3*	a = 0·8*
Minneapolis	45°N	325	8 382	74 423	70 651	20 452	19 335	70 560	62 229	19 378	16 787	66 066	51 298	18 109	13 530
Concord, N.H.	43°N	300	7 000	68 759	64 826	18 895	17 714	64 674	55 363	17 743	14 972	59 759	43 667	16 370	11 344
Denver	40°N	425	6 283	57 337	53 943	15 755	14 824	53 726	44 937	14 763	12 198	48 780	34 095	13,405	8 720
Chicago	42°N	350	6 155	58 516	55 356	16 081	15 210	55 219	47 678	15 169	12 865	50 684	37 339	13 847	9 743
St. Louis	39°N	375	4 900	45 046	42 149	12 379	11 565	41 981	35 192	11 533	9 476	38 038	26 344	10 425	6 660
New York	41°N	350	4 871	45 906	42 950	12 615	11 804	42 843	35 368	11 774	9 594	38 385	25 231	10 548	6 406
San Francisco	38°N	410	3 015	23 258	21 120	6 392	5 803	20 916	15 631	5 748	4 118	16 948	9 812	4 645	1 743
Atlanta	34°N	390	2 983	26 922	24 803	7 398	6 771	24 475	19 206	6 716	5 103	20 639	12 399	5 562	2 587
Los Angeles	34°N	470	2 061	9 900	8 549	2 720	2 349	8 392	5 758	2 306	1 316	6 139	3 040	1 520	155
Phoenix	33°N	520	1 765	11 861	10 533	3 259	2 878	10 283	7 316	2 826	1 811	8 077	4 619	2 062	555
Houston	30°N	430	1 600	14 592	12 956	4 011	3 557	12 888	9 379	3 542	2 351	10 878	6 760	2 909	1 142
Miami	26°N	451	141	210	106	7	0	92	0	0	0	0	0	0	0

* 'a' is absorptivity or absorption coefficient

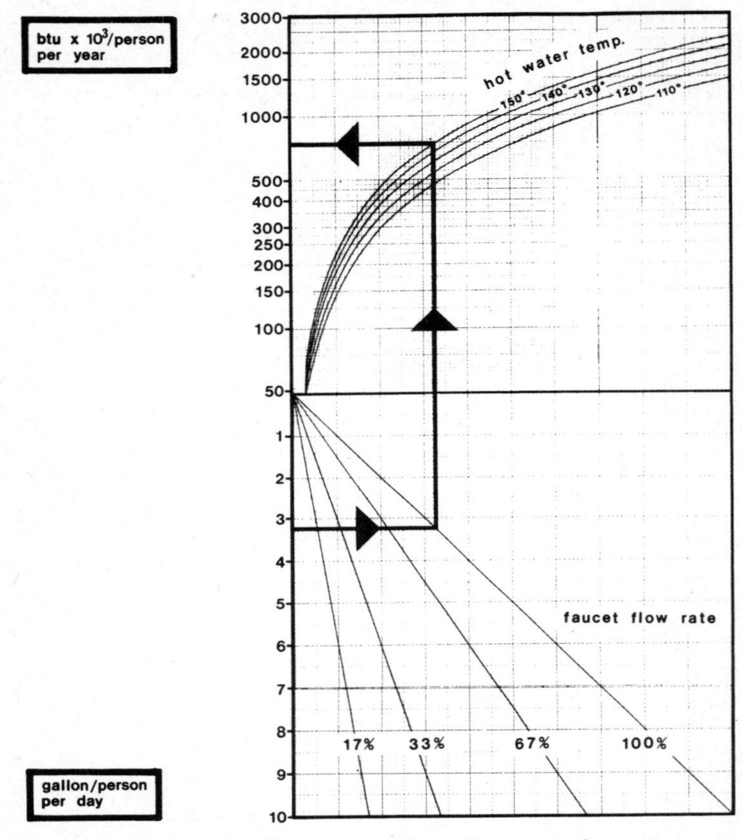

Fig. 3. Hot water: saving for reduction of faucet flow rate and water temperature.

savings of from 40 % to 50 % are possible in new buildings compared with current usage.

But energy conservation does not 'just happen' by itself. Nor do turning off the lights and turning down the thermostats constitute an adequate energy conservation plan. Instead, we have found that a thorough study must be undertaken to determine the methods for saving energy, the amount that can be saved, both generally for building types, and specifically for each building, and a procedure must then be worked out for implementing changes to be made as a result of the in-depth study.

As a professional consulting engineering and planning firm, we have been commissioned on three separate occasions by US Government agencies to conduct studies for the purpose of planning energy conservation methods in buildings. These studies have broadened our understanding of the immediate and potential value of energy conservation studies. From our experience we have found—in the process of conducting the studies—that:

Each energy conservation study, though limited by time and budget constraints, discloses additional areas for investigation.
Each study becomes a building block in the growing structure of conservation practices, and becomes part of the body of knowledge which is vital to a national energy policy.
The unsuspected and hitherto often neglected inter-relationship between climate, structure, mechanical and electrical systems, and building use and operation which affect energy consumption are brought into focus and quantified.
The potential for conservation becomes more clearly apparent with each study, and realistic goals for limits of energy use (such as energy budgets in MJ/m^2 yr (Btu/ft^2 yr) or other comparable standards) can be assigned individually for each building type in each climatic zone in the United States.
Ultimately, the options and guidelines which emanate from energy conservation studies can be, and are being, used as educational material, text books and reference for students and trainee engineers, architects and planners.
Energy conservation studies disclose that new and existing buildings can be treated alike to a greater extent than was commonly believed.

In addition, we have drawn some general conclusions about the value of energy conservation studies. Energy Conservation Studies:

provide technical data and procedures which are of value to municipal and federal legislators, and code officials who are considering energy conservation legislation.
provide a stimulus to architects and engineers to further address energy conservation problems in new and existing buildings, and educate them into the methods of preparing and implementing energy conservation programmes.
provide information which enable the formulators and designers of alternative energy sources to address the prerequisites required of

the building to optimise such systems as solar and total energy systems.
provide opportunities to examine the many and varied computer programmes and analytical methods which are available for load analyses and energy supply systems design.
which include energy management programmes are of value to professionals called upon to address load management, as power plant facilities become increasingly over-burdened.
provide methods to quantify measures which have been previously qualified only.
provide methods for performing economic and life cycle cost analyses for energy conservation measures on both an absolute and comparative basis.
of a single building or building type are also of generic interest and of greater value when, at a minimum, the sensitivity to other climates and diverse variables are also examined.

Energy Conservation Demonstration Project, Federal Office Building, Manchester, New Hampshire

In 1972, at a conference sponsored by the Federal Government's General Services Administration (GSA) and the National Bureau of Standards (NBS), the author (FSB) suggested that the time had come to test, in an operating building, the myriad energy conservation ideas which were discussed in seminars and speeches and described in magazine articles which had virtually inundated the ears, eyes, and offices of engineers, architects, government officials, legislators, and the public at large. On October 2, 1972, the new Federal Office Building to be constructed in Manchester, New Hampshire, was designated by Arthur F. Sampson, Administrator of General Services, as the GSA Energy Conservation Demonstration Building. The purpose of this project, stated the GSA 'is to (a) dramatise the firm commitment of PBS/GSA to the conservation of energy in the design, construction and operation of federal buildings; (b) provide a laboratory for the installation of both recognised and innovative energy conservation techniques/equipment; and (c) inspire others in the building construction industry to pursue energy conservation as a goal'. A professional services contract was awarded to Nicholas Isaak

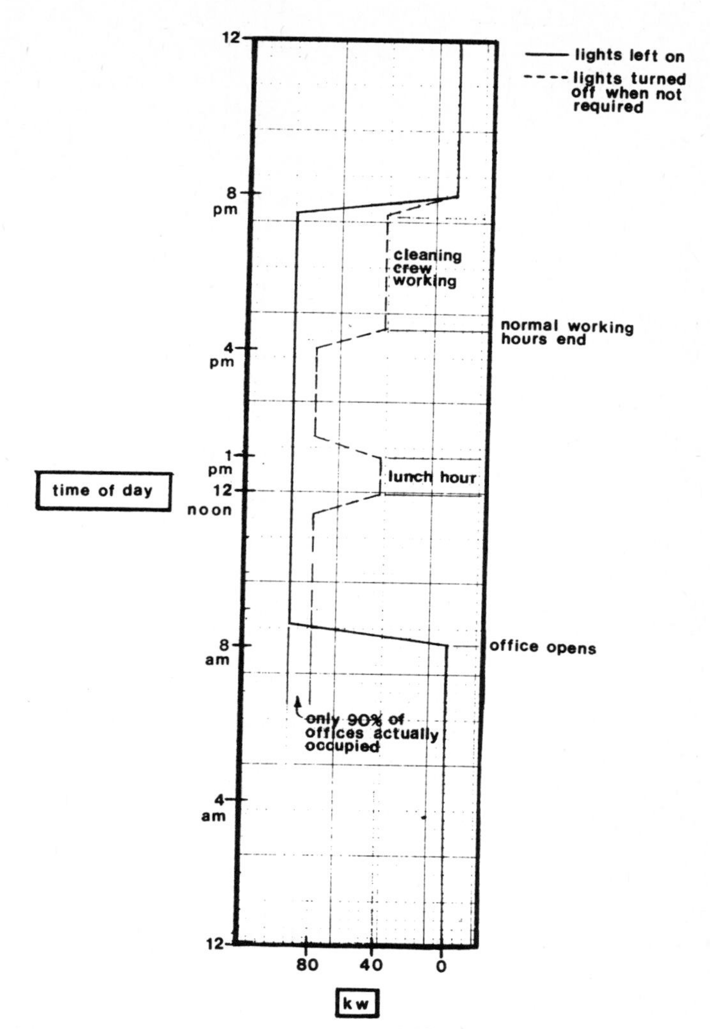

Fig. 4. Lighting: the effect of turning off unnecessary lights on power consumption. Example: assume an office lighting system of 1000 *two-lamp fluorescent luminaires with f*40 *lamps. The full system uses* 92 k *W. The solid line shows typical usage when lights are permitted to burn continuously from about* 8.30 *a.m. to* 7.30 *p.m. The dashed line shows efficient use of the same system leaving lights on only when used. Energy saving amounts to* 262·5 k *Wh daily or* 5875 k *Wh per month.*

and Andrew C. Isaak, Architects, Manchester, New Hampshire; R. D. Kimball Associates, Engineers, Boston, Massachusetts for the design of the project, and DMBA to serve as Energy Conservation Consultants for this project. NBS was engaged to perform detailed computer studies and to participate in the instrumentation and subsequent evaluation of the building. DMBA was responsible for the collection, evaluation, and selection (with GSA concurrence) of energy conservation features to be included in the project.

In the opening phase of the study, we solicited ideas from utility companies, manufacturers, designers, and universities, among others. The result was some 500 suggestions. We analysed many of them in detail and rejected some out of hand. We also reviewed the details of 7000 of our earlier building projects, re-examining them for energy saving ideas and analyses which we had previously performed.

For analysis, we first established a base building with standard construction. The vital statistics of a 'Typical Design Practice Building' (an ordinary non-conservation oriented office building) devised as a control model for computerised comparative analysis were agreed upon by representatives of GSA, DMBA and NBS as follows:

A six-storey building, with 1950 m^2 (21 000 ft^2) of space per floor, rendering a total net office space of 11 700 m^2 (126 000 ft^2), with an additional area of 3900 m^2 (42 000 ft^2) for underground parking garage was developed. This building had a two-to-one length-to-width aspect ratio with the long axis running north to south. The U value (the coefficient of heat transmission) established was to be not greater than 1·87 W/m^2 °C (0·33 Btu/h ft^2 °F) for exterior wall construction; the U value for roof construction of not greater than 0·85 W/m^2 °C (0·15 Btu/h ft^2 °F), single pane windows with an inside shading coefficient of 0·50, and a winter U value of 6·42 W/m^2 °C (1·13 Btu/h ft^2 °F). The initial non-energy conservation building could have had an exterior wall of conventional masonry, or curtain wall construction, using lightly insulated panels. The glass area would have covered 50 % of the wall surface. The building would not have exterior shading devices of any kind.

Mechanical and electrical factors to complete the mathematical model were also set:

(a) Wattage for lighting was to be 36·6 W/m^2 (3·4 W/ft^2) for 75 % of the floor area, and 10·8 W/m^2 (1 W/ft^2) for 25 % of the area.

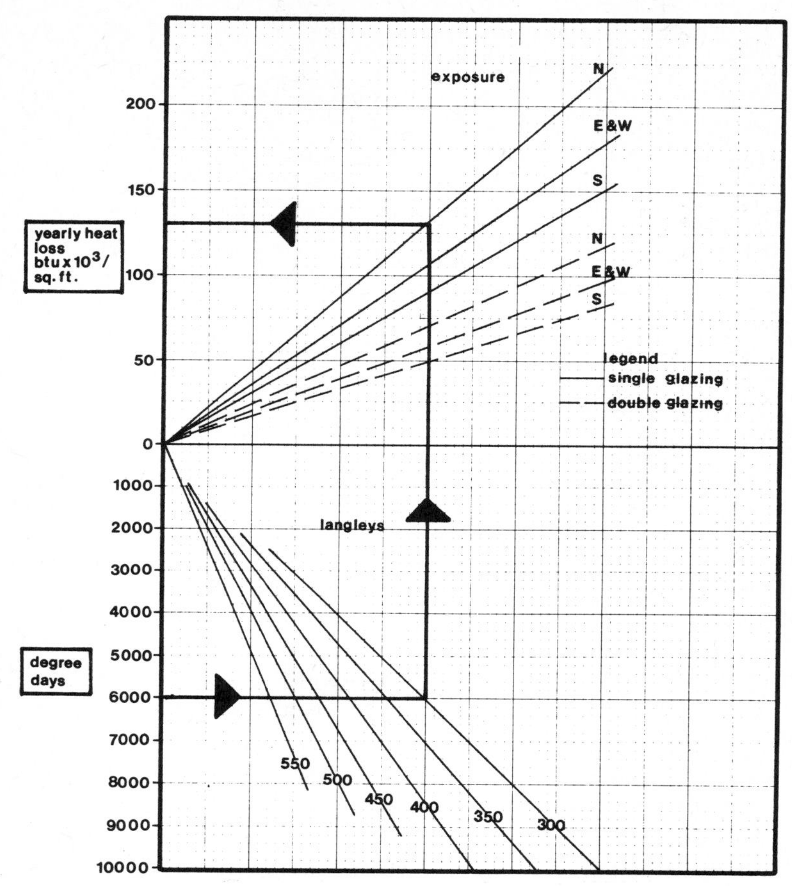

Fig. 5. Heating: yearly heat loss through windows—latitudes 35°N–45°N.

(b) Office equipment was selected to use 5·4 W/m² (0·5 W/ft²) for 75% of the floor area.

(c) Based on occupancy of 600 people, outdoor ventilation air was established at 7·08 m³/s (15 000 ft³/m). Infiltration air was assumed at 0·25 air change per hour.

Many of these criteria were adopted to satisfy various building codes and presupposed the non-existence of energy recovery systems.

The NBS computer programme (NBSLD) was used to determine the building load for heating, ventilating, domestic hot water, air conditioning, illumination and power systems. The output from this programme was used as input to three other computer programmes to calculate the amount of energy required each month for each of the environmental control systems. Heat gains and losses were determined as functions of different building components—roof, wall, orientation, height, floor, among others. Some of the parameters included:

Thermal resistance and mass of walls
Window-to-wall area ratio
Single, double, triple-glazing
Solar shading
Building orientation, configuration and length
Length to width ratio
Exterior wall colour
Insulation values and location on the inside and/or the outside of the exterior wall
Minimum air infiltration
Use of natural lighting to minimise electric power consumption
Hours of heating and cooling systems operation during unoccupied periods

The typical design practice building would have had a heat loss of about 1172 kW (4×10^6 Btu/h), and a peak cooling load of about 1055 kW (300 tons refrigeration). Complete yearly analysis on an hour-by-hour basis was made of dynamic energy flow using tapes of actual weather data for the New Hampshire area.

Built to current standards of construction, the building would have required an input of about 1317 MJ/m^2 yr (116 000 Btu/ft^2 yr) at the building, and about 2453 MJ/m^2 yr (216 000 Btu/ft^2 yr) of primary source energy.* Similar size buildings consume as much as 6810 MJ/m^2 yr (600 000 Btu/ft^2 yr) in the New England area.

Using the same computer programmes and the parameters listed above, more than 75 computer runs were made to determine the optimum building envelope. Each of the variables was changed parametrically, with all other variables held constant, so that potential savings which were attributable to each sub-component could be determined.

* Includes the energy to generate electricity, assuming an efficiency of 30 % for generation and electrical transmission lines.

More than 25 computer runs were then made to analyse annual energy use for a wide number of heating, ventilating, cooling and lighting systems, individually, and in combination with each other.

The final analyses of the building and systems showed a potential energy saving of more than 50 %. These savings are achieved by using available off-the-shelf hardware as part of the requirements of the study to avoid recommending custom-made or exotic components.

From these overall analyses, design criteria were developed for a building that would use 40 % to 50 % less energy than a building designed to current standards and practices. Computerised analyses indicate that the finished building will use only 624 MJ/m^2 yr (55 000 Btu/ft^2 yr) at the building boundary and less than 1250 MJ/m^2 yr (110 000 Btu/ft^2 yr) of raw source of energy—a reduction in energy use of more than 50 %.

In addition to the annual reduction in energy use, the heat loss at design conditions was reduced to 615 kW (2 100 000 Btu/h) and the peak design cooling load to less than 527 kW (150 tons) of refrigeration. The savings in capital costs for these systems, and the reduction in peak electrical demand with corresponding savings in electricity costs contribute significantly to the lower life cycle costs which will accrue.

After a subsequent study, which is described below, an energy budget of 624 MJ/m^2 yr (55 000 Btu/ft^2 yr) was established as a target goal by GSA for all new federal office buildings designed for them. This goal has been met in the four office buildings designed this year for GSA in four distinct climatic zones in the USA. Other federal agencies, and many private agencies, have adopted the same energy budget for their new buildings.

Before describing the next study, it may be of interest to briefly note some of the features which were recommended and adopted for the Manchester building which is now under construction.

SELECTED FEATURES OF THE GSA DEMONSTRATION BUILDING

Building envelope:

Configuration: 7 storeys with additional underground garage levels; typical floor dimensions, 41 m × 30 m (135 ft × 100 ft).

Roof: Mass, 390 kg/m^2 (80 lb/ft^2); U Value 0·34 W/m^2 °C (0·06 Btu/h ft^2 °F); light reflective.

Exterior Walls: Mass, 390 kg/m^2 (80 lb/ft^2); U Value 0·34 W/m^2 °C

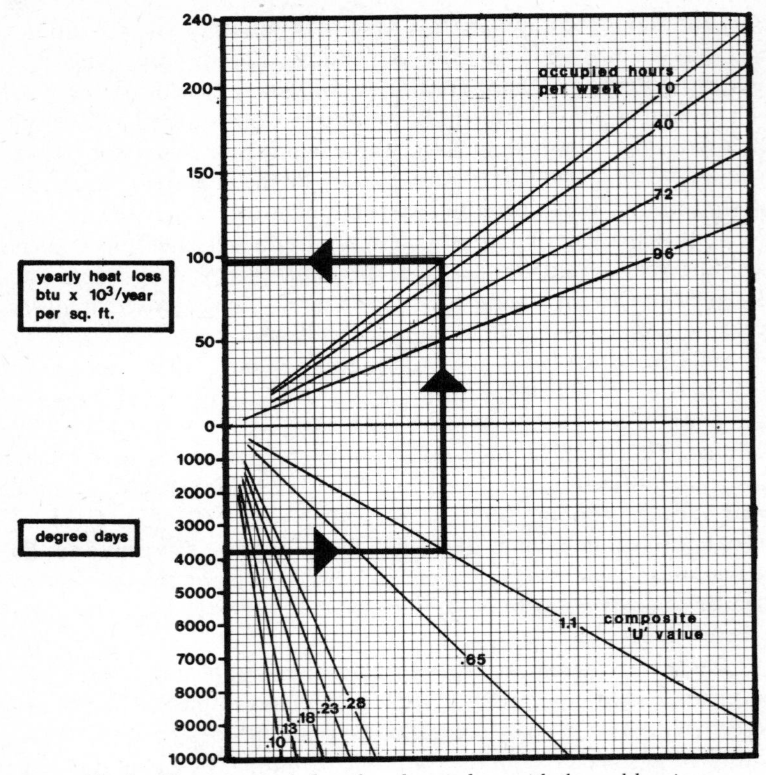

Fig. 6. Heating: yearly heat loss for windows with thermal barriers.

(0·06 Btu/h ft^2 °F); insulation location outside surface of exterior wall.

Windows: North facade, none; window to wall area for 6 floors; 12·5 %; window to wall area for naturally illuminated floor, 30 %; double glazed, tinted; infiltration rate, less than $0{\cdot}3 \times 10^{-3}$ m^3/s/m of crack (0·2 ft^3/min/ft of crack).

Solar Control: South facade, mainly horizontal overhangs with small vertical exterior fins; east and west facades, major vertical exterior fins and small horizontal overhangs.

Building plan:

Interiors: Open landscape planning; light reflective walls, floors and ceiling; buffer spaces on the north wall are corridors, storage areas, lobby and mechanical rooms.

HVAC systems:

Environmental Conditions: Ventilation, 2·8 litres/s (6 ft^3/m) per person during occupied hours; temperature levels for heating, maximum of 20 °C (68 °F) during occupied periods with lower temperatures in non-critical areas, setback of 6 °C (10 °F) during unoccupied hours; temperature levels for cooling 25·6 °C (78 °F) with higher temperatures in non-critical areas and shut-down during unoccupied hours; relative humidity, summer uncontrolled below 60 % RH. and winter uncontrolled above 20 % RH.

Zoning: Dampers, switches, and controls to permit selective and variable conditions.

Systems: Different systems on each floor to permit evaluation of relative performance, including closed-loop heat pumps; engine driven emergency generator powered heat pump; variable air volume in the interior and exterior zones; fan coil systems which can operate as 2, 3 or 4 pipe systems; separate air handling units for interior and individually for each of the four exterior perimeter zones; blow-through air handling units; compressors piped in series arrangements; enthalpy control with economiser cycle; recirculated toilet exhaust air through charcoal filters; air filters with low frictional resistance; electric ignition instead of gas pilots; CO_2 monitoring system in garages; hot water and chilled water storage tanks.

Light and power:

Lighting: Photo cell control for natural light floor; selective lighting on all floors to include lighting and low level ballasts; average power for lighting, 20·5 W/m^2 (1·9 W/ft^2); furniture mounted task lights on first floor; multi-level ballasts; polarised lighting; high output lamp sources including sodium on one floor; mercury vapour lamps in garages.

Solar energy systems:

The solar energy system includes approximately 437 m^2 (4700 ft^2) of solar collectors, mounted on the roof in four rows. The collector performance and space requirements are based on flat plate collectors, facing south and arranged with a mechanism to provide adjustable tilt from 20° to 80°. The angle adjustment will be done seasonally to provide the amount of heat and temperature required for heating and cooling—the lower angles in the summer are best for cooling.

The three storage tanks are used to store hot water from solar collectors, or waste heat from the diesel-gas engine and chilled water in the summer time. Heating coils were selected to permit operation with hot water at low temperatures to increase the collector efficiency. The heat pumps will boost the water temperature from the collector to increase seasonal efficiency.

We found our work on the Manchester study to be important in helping us to determine energy-use relationships, and realistic energy budgets. While in essence the study was for a specific single building, in another sense, it was generic in that the body of knowledge resulting from this building study could be transferred to other buildings in other areas.

GSA Energy Conservation Guidelines for New Office Buildings:

From the Manchester study grew the realisation that guidelines for all GSA buildings, regardless of climate, should be developed. After our experience on the Manchester project, we were commissioned to do guidelines for all new structures and were assisted in this effort by Heery and Heery, architects, and the AIA Research Corp. These guidelines for new buildings include a greater number of conservation options and are generally more comprehensive.

GSA buildings, located throughout the United States, range from 1 to 25 storeys in height and cover up to 92 900 m^2 (1×10^6 ft^2) of space each.

Energy usage in office buildings was identified by sub-system, even to the extent of the thermal dynamics of individual exterior building walls, separately for each exposure, in various geographic climatic zones. Energy usage and conservation measures are included for user needs related to the amount of hot water, temperature, humidity and ventilation air required; the amount and quality of illumination.

The energy use attributable to site development, building envelope, structure and plan, building use, type and efficiency of mechanical and electrical systems, and the impact of each factor on the other was identified.

The guidelines describe various methods that can be used to reduce energy consumption for new buildings by 30 to 60% below current practices. They have been arranged to permit architects and engineers the greatest amount of design latitude, and to encourage innovation in design which is responsive to energy conservation principles. And they are based on methods, which, if followed, will make it possible to meet the energy budget goal of 625 MJ/m^2 yr (55 000 Btu/ft^2 yr) of energy input at the building boundary and 1249 MJ/m^2 yr (110 000 Btu/ft^2 yr) raw source energy regardless of location. While these guidelines were developed for new buildings, many of the individual measures can be applied to retrofit of existing buildings as well. GSA has since produced another manual for existing buildings using much of the data and many of the guidelines which were developed for new buildings.

The GSA manual[1] is based on the concept that to design for energy conservation, the entire building must be considered as a whole—a system in which there is interaction between the building envelope, the structure, the mechanical and electrical sub-systems and the users of the building. The energy conservation measures are listed in order of priority in accordance with their sensitivity to climate and location in the last section of the manual. However, they are not prescriptive, except for the energy budget goal.

The additional studies, described below, which DMBA performed for the Federal Energy Administration for existing buildings, advance the concept of conservation studies another step by quantifying the potential energy and dollar savings for each of the many energy conservation opportunities which were studied and detailed.

Energy Conservation Studies for Existing Buildings for FEA

The US Government's Federal Energy Administration (FEA) has the responsibility to formulate policy and implement measures to reduce the gap between domestic energy supplies and demand. To assist them in their efforts to stimulate energy conservation measures in buildings, DMBA was commissioned in June 1974 to study energy use in existing office, retail and religious buildings; research and develop energy conservation measures; and prepare a set of manuals for public use detailing the energy conservation options which are available. However, as the studies progressed, it became increasingly apparent that most of the conservation measures were applicable to university and hospital buildings, apartment houses, schools and industrial

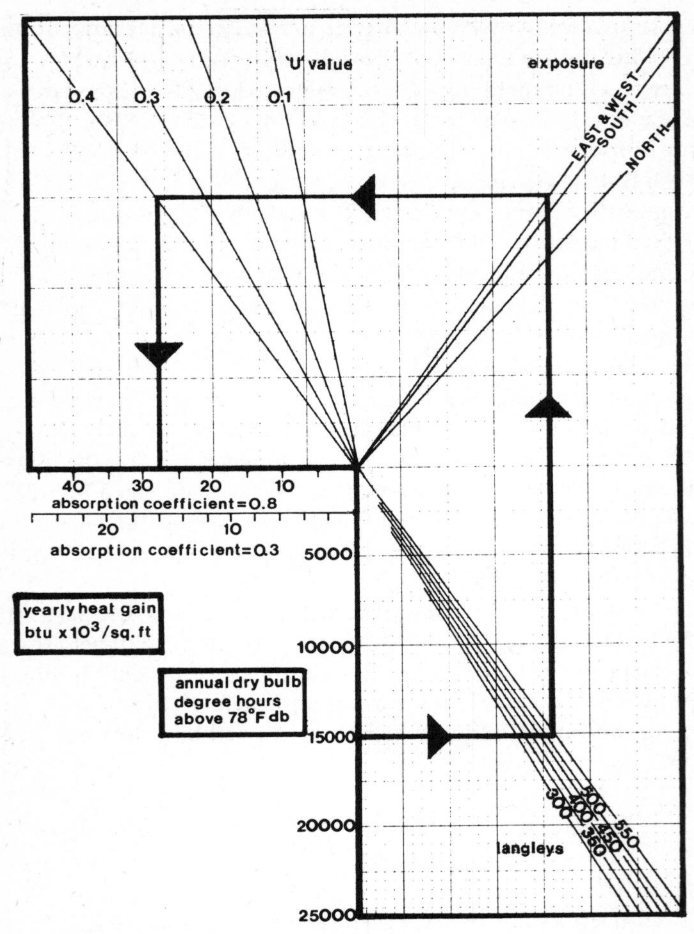

Fig. 7. Cooling: yearly solar heat gain through walls—latitudes 35°*N*–45°*N.*

plants (excluding industrial process) as well; and more than 80% of the same energy conservation options were applicable to new buildings. Based on the studies, DMBA prepared two manuals[2] entitled 'Energy Conservation for Existing Buildings, Opportunities and Guidelines'—'ECM–1' and 'ECM–2'. ECM–1 is designed to help building owners, managers, operators or occupants to conserve and manage energy usage now, without the need of investing any significant amount of money to do so. It is not strictly a design manual, nor a cost estimating manual, but rather a set of guidelines and procedures to save energy by reducing waste, and by more effective operation of the building and its mechanical and electrical systems. ECM–1 provides case histories of energy and cost savings for two buildings which have already instituted energy conservation programmes. These two buildings were able to save $758 000 and $434 000 per year respectively, with little capital costs, by following the recommendations of professional engineers retained by the owners. Many of the ideas outlined in these guidelines can be executed directly by the building owners, managers or occupants without delay or the need for further advice, while other measures, even though they do not entail significant capital costs may require further analysis by an engineer, architect, utility company or service and maintenance organisation.

The manual contains a summary of the major procedures that can be followed to save energy and operating costs for the building. GSA, which operates more than 10 000 buildings reports an average savings of 30 % in energy by following many of these procedures. A feature of the manual is the profile forms which permit an owner to identify his building by climate, use, construction type, and mechanical and electrical systems and characteristics. Using the curves and charts in the manual, an experienced investigator can quantify the potential savings in energy for each energy conservation measure. (See Figures 1 to 14 and Tables 1 and 2 for examples.)

ECM–2 was prepared as a working tool for architects and engineers who analyse energy usage and recommend and design appropriate measures to conserve energy in existing buildings, as well as for building operating personnel who are responsible for an on-going energy management programme and the maintenance and operation of the building and its environmental control systems.

The energy conservation options detailed in ECM–2 include measures to:

(1) Maintain indoor environmental conditions at various and different levels in the same building at the same time.
(2) Reduce the heating, cooling and lighting loads by modifying the building structure or site.
(3) Improve the seasonal efficiency of the primary energy-conversion equipment and distribution systems by modification or replacement.

ECM–2 also provides the elements of an energy management programme with detailed instructions for:

conducting an energy audit.
determining the potential for energy conservation from a comprehensive list of energy conservation measures.
calculating the order-of-magnitude of costs for equipment, systems and materials.
determining short and long-term economic feasibility of any energy conservation option by using an economic model.
providing methods of shedding loads to reduce peak electricity demand in order to reduce demand charges and conserve energy.
developing heat reclamation opportunities and provisions for present and future installation of solar energy and total energy systems. Also included are computer programme sources and applicable codes and standards.

We have extended the principles which we explored, and the knowledge gained from the three government-sponsored studies, to the private sector and other governmental and municipal agencies, who have retained us for studies and application to their existing and proposed new buildings.

It has become increasingly apparent that even though all buildings have much in common, and that the principles of energy conservation for buildings are univeral, each individual structure is unique, even though types may be similar and even located immediately adjacent to each other. A conservation analysis must be performed for each building individually. A particular energy conservation measure which is effective in one building may be counter-productive in another.

Some of the additional public and commercial projects are briefly described below.

Connecticut General Life Insurance Building, Bloomefield, Connecticut

The government studies provided a valuable data base to extend energy conservation studies and designs to conserve energy for this existing building. In 1973, DMBA analysed the loads due to the building envelope and Heating Ventilating Air Conditioning (HVAC) distribution systems, and made an energy audit of the fuel and electricity required for heating, ventilating, air conditioning, lighting, illumination and power. The building and mechanical and electrical system operations were studied at length with the building engineer, and with the building management. DMBA prepared an energy management programme which could be implemented in increments, for management's approval. The cost/benefits of alternative measures were documented, but in some cases, the savings in costs were so clearly apparent that management authorised DMBA to proceed with design without completing detailed costs/benefits in order to take immediate advantage of conservation opportunities in the face of rising energy costs. Features of the main building, prior to the conservation programme, are summarised below.

STRUCTURE:

Three storeys; rectangular plan with six internal courts; 83 600 m^2 (900 000 ft^2) gross floor area; insulated curtain wall construction with single thickness heat absorbing glass in upper panels and clear plate glass in lower panels; vertical slat venetian blinds.

MECHANICAL AND ELECTRICAL SYSTEMS:

Two water tube steam boilers 3·5 kg/s (27 500 lb/h) capacity, operating pressure 1·4 MPa (200 lb/in^2); system designed to operate with coal, No. 6 oil, and gas; systems operating with No. 6 oil with forced and induced draft fans sized for coal burning; steam distribution to two steam driven 2·8 MW (800 ton) chillers; 310 kPa (45 lb/in^2) and 172 kPa (25 lb/in^2) steam to kitchens for dishwashing and boosting domestic hot water; 35 kPa (5 lb/in^2) steam to air heating coils, heat exchangers for hot water heating elements, and domestic hot water; perimeter induction units serving 13 zones; low pressure air handling units serving interior zones; return air from perimeter zones and interior zones returned to the appropriate units for the major portions of the building but mixed in about 25 % of the building; 9 of the HVAC units include spray cooling coils and steam humidifiers; perimeter air handling units designed to supply air at temperatures of 8·9—48·9 °C (48—120 °F); 10 deep wells provide cooling for the minimum outdoor air setting in precooling coils, condenser water cooling for the two 2·8 MW (800 ton) chillers, and condensing media for the turbines; one electric 3·2 MW (900 ton) centrifugal chiller operating with cooling tower; precooling coils using

TABLE 2a

YEARLY HEAT LOSS/SQUARE METRE THROUGH ROOF

City	*Latitude*	*Solar radiation*	*Celsius base*	*Heat loss through roof* MJ/m^2 *yr*			
				U = 1·08 W/m^2 °C		U = 0·68 W/m^2 °C	
		MJ/m^2 *day*	*degree days*	a = 0·3*	a = 0·8*	a = 0·3*	a = 0·8*
Minneapolis	45 °N	13·6	4657	400·3	351·7	242·2	211·7
Concord, N.H	43 °N	12·6	3889	368·6	314·3	223·1	188·8
Denver	40 °N	17·8	3491	304·3	255·3	184·3	153.3
Chicago	42 °N	14·7	3419	312.2	267·9	188·9	161·1
St. Louis	39 °N	15·7	2722	238·2	198·0	144·1	118.7
New York	41 °N	14·7	2706	242·2	196·7	146·6	118·3
San Francisco	38 °N	17·2	1675	119·8	91·9	72·5	54·3
Atlanta	34 °N	16·3	1657	143·1	111·8	86·5	66·2
Los Angeles	34 °N	19·7	1145	52·6	42·0	31·7	24·3
Phoenix	33 °N	21·8	981	65·8	53·6	39·6	31·3
Houston	30 °N	18·0	889	68·6	54·5	41·1	31·5
Miami	26 °N	18·9	78	2·94	1·48	1·58	0·62

* 'a' is absorptivity or absorption coefficient

TABLE 2b
YEARLY HEAT LOSS/SQUARE FOOT THROUGH ROOF

City	*Latitude*	*Solar radiation Langley's*	*Degree days*	*Heat loss through roof Btu/ft² year* U = 0·19		U = 0·12	
				a = 0·3*	a = 0·8*	a = 0·3*	a = 0·8*
Minneapolis	45 °N	325	8 382	35 250	30 967	21 330	18 642
Concord, N.H.	43 °N	300	7 000	32 462	27 678	19 649	16 625
Denver	40 °N	425	6 283	26 794	22 483	16 226	13 496
Chicago	42 °N	350	6 155	27 489	23 590	16 633	14 190
St. Louis	39 °N	375	4 900	20 975	17 438	12 692	10 457
New York	41 °N	350	4 871	21 325	17,325	12 911	10 416
San Francisco	38 °N	410	3 015	10 551	8 091	6 381	4 784
Atlanta	34 °N	390	2 983	12 601	9 841	7 619	5 832
Los Angeles	34 °N	470	2 061	4 632	3 696	2 790	2 142
Phoenix	33 °N	520	1 765	5 791	4 723	3 487	2 756
Houston	30 °N	430	1 600	6 045	4 796	3 616	2 778
Miami	26 °N	451	141	259	130	139	55

* 'a' is absorptivity or absorption coefficient

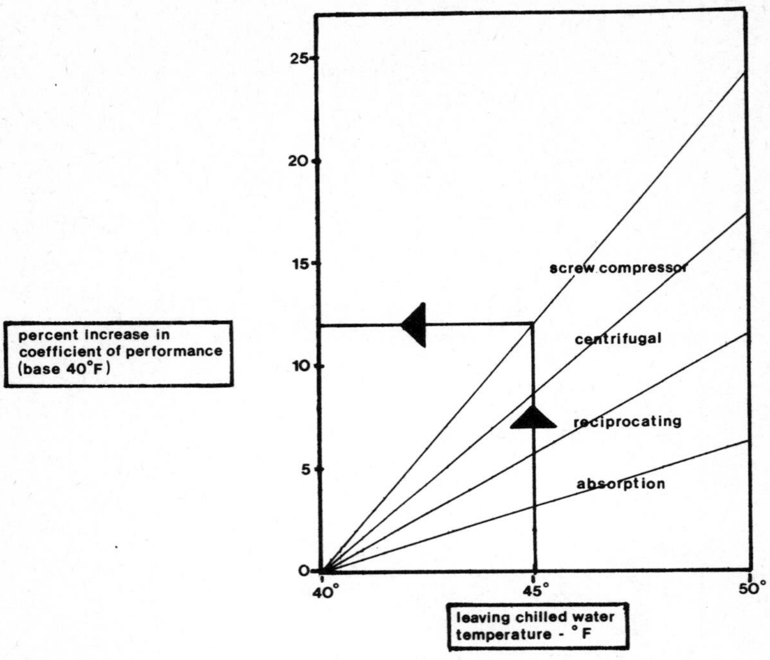

Fig. 8. Cooling: effect of chilled water temperature on chiller coefficient of performance—c.o.p.

well water provide 2·5 MW (700 tons) of refrigeration. Total air handled approximately 470 m^3/s (1×10^6 ft^3/min).

FEATURES OF THE PRINT SHOP ADDITION BEFORE CONSERVATION:

The print shop is in a separate building and consists of 9300 m^2 (100 000 ft^2) gross area with separate electric heating and cooling systems.

MANAGEMENTS'S ATTITUDE AND ENERGY CONSERVATION OBJECTIVES AND REQUIREMENTS:

The employees (building tenants) must recognise and moderate their comfort level requirements, *i.e.*, 20—25·6 °C (68—78 °F); conserve natural resources in line with CG's social commitments; as energy costs rise and erratic inflation governs, operating monies must be efficiently used; employee productivity must increase rather than drop as a result of energy conservation.

BASIS OF DECISIONS:

Life cycle costs (payback period) are to be considered; aesthetic impact on the buildings not to be impaired; employee reactions; initial capital investment involved; efficiency of operation resulting from decision to invest capital in energy conservation; savings that will occur in manpower, *i.e.*, not adding additional manpower and/or not replacing positions caused by retirements.

INVESTMENT LIMITS:

'Payback' periods are generally considered to be four years or less; corporation cash flow situation at the time an investment is to be made; impact of investments on the rental cost to tenants of the building.

CAPITAL INVESTMENT CONSIDERATION:

Forecasting of energy costs on a one plus one basis using best corporate information available at the time of consideration; cost of rent to the tenant versus other costs of doing business for the tenant; fiscal taxable considerations as delineated by CG Tax Department.

Energy Conservation Measures Which Have Been Implemented

Boilers converted for firing No. 2 oil or natural gas with new steam atomising burners to reduce air pollution resulting from the high sulphuric contents of No. 6 oil available in the area; boilers de-rated to steam capacity of 2·8 kg/s (22 000 lb/h); draft fans reduced in speed and fitted with inlet vortex damper controlled by flue gas oxygen analyser; burners fitted with automatic turn-down ratio of 4 to 1 replace inefficient combustion control system; all boiler output control at lower capacities is manual at present; (the original control system provided for automatic change over from heating to cooling at an outdoor temperature of 4·4 °C (40 °F) for the perimeter zones, and energy was wasted because the system operated either with reheat or with recooling;) the induction changeover controls were disconnected and hot water or chilled water to the coils programmed manually (experience now shows that secondary cooling is not required now until outdoor temperature rises to 15·6 °C (60 °F)); space temperatures during occupied periods have been reduced from 23·9 °C (75 °F) in the winter to 20 °C (68 °F) when heating is required and raised from 23·9—25·6 °C (75—78 °F) in the summer when cooling is required; no reheat is used; ventilation requirements reduced from 7—2·8 litres/s per person (15—6 ft^3/min); all down escalators turned off during the day except for heavy traffic periods; lighting which had been uniform at 600 lx (60 ft candles) was reduced in corridors, cafeterias, lobbies,

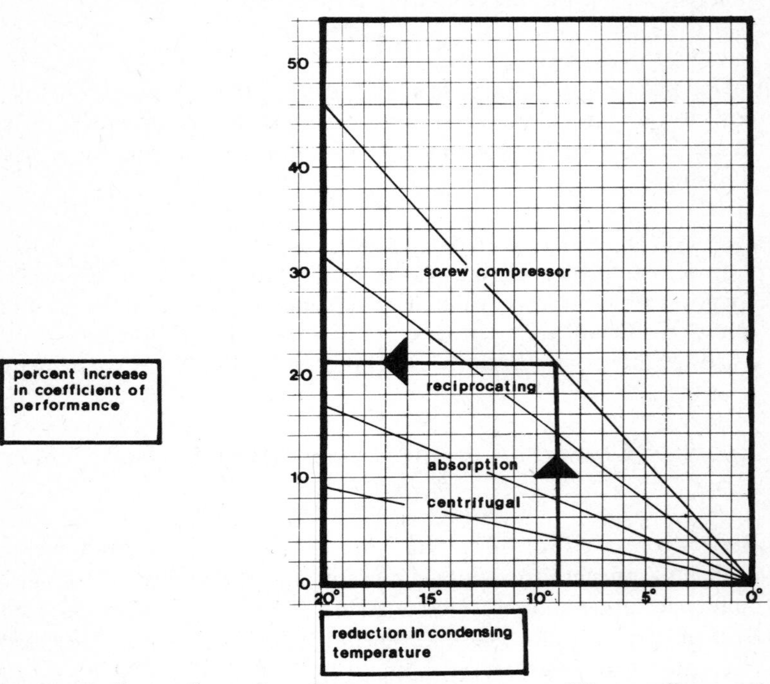

Fig. 9. Cooling: effect of condenser temperature on chiller coefficient of performance—c.o.p.

equipment rooms and in other areas with less demanding tasks by removing lighting bulbs and disconnecting ballasts; when relamping occurs, lamps with greater efficiency (more lumens per watt) are installed; one additional deep well installed; spray coils in 9 HVAC units modified to provide adiabatic cooling and spray coil humidification; one of four additional units has been similarly equipped and is under test before completing modifications of all units; modifications made to permit cooling with well water for return air and outdoor air in place of cooling minimum outdoor air only, and cooling load reduced by about 2·5 MW (700 tons) enabling centrifugal units to be removed from service for a large percentage of the time, thus reducing both electrical consumption and electrical demand; the additional well provides sufficient capacity for condenser cooling to replace the cooling tower for the 3·2 MW (900 ton) chiller when in operation; enthalpy controller has been installed on one air handling unit for testing and evaluation; domestic hot water temperature reduced from 60—38·8 °C (140—100 °F), but raised to 48·9 °C (120 °F) when complaints arose.

The additional measures which were considered and either rejected or the implementation delayed are outlined below; but first the results of the present programme are detailed:

(1) The total consumption of energy for the main building in 1972 prior to conservation was 213 TJ (201 224 × 10^6 Btu) the equivalent of 2539 MJ/m^2 yr (223 580 Btu/ft^2 yr) for the main building.

(2) After the first year of conservation, the energy consumption for the main building in 1973 was reduced to 199 TJ/yr (188 323 × 10^6 Btu/yr), but an extra load of 5374 GJ/yr (544 × 10^6 Btu/yr) was added for processing. The gross consumption including the added process load was the equivalent of 2383 MJ/m^2 yr (209 850 Btu/ft^2 yr).

(3) In 1974, with the continuing conservation programme, the yearly consumption for the main building dropped to 173 TJ (164 138 × 10^6 Btu) but extra process load and the print shop loads raised the total to 174 TJ (165 131 × 10^6 Btu/yr) or the equivalent of 1875 MJ/m^2 yr (165 130 Btu/ft^2 yr). The actual savings in energy conservation amounted to 722 MJ/m^2 yr (63,580 Btu/ft^2 yr) or a reduction of 29 %.

Because of rising fuel prices, and the addition of the new loads in 1973 and 1974, the operating costs for energy increased 37 %. However, it is important to keep in mind that without the conservation programme the energy costs would have been 13 % higher at 1974 fuel costs, and considerably higher than that at current fuel prices (1975).

ADDITIONAL ACTIONS TAKEN:

A new computerised control system has been installed for load levelling and optimisation of the HVAC and illumination systems to

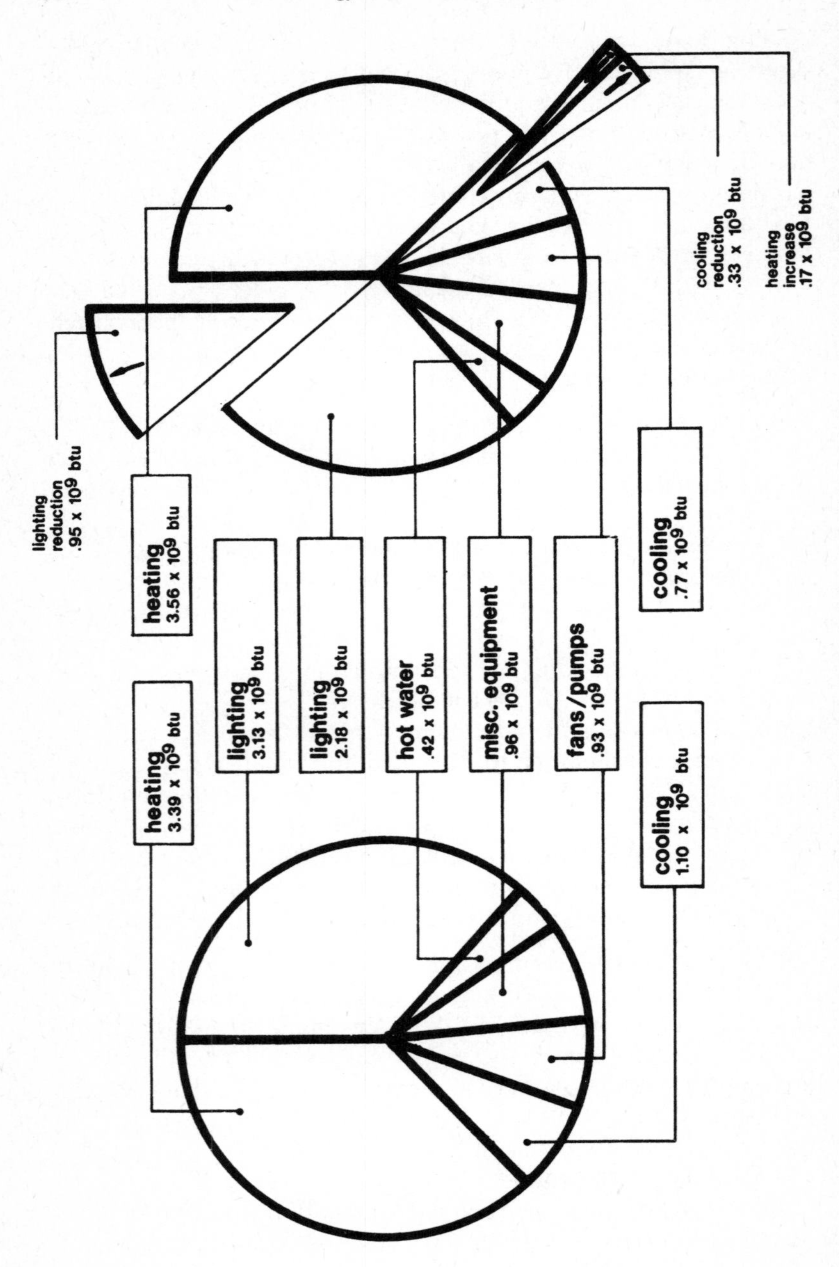
heating
3.39 x 10^9 btu
lighting
3.13 x 10^9 btu
lighting
2.18 x 10^9 btu
hot water
.42 x 10^9 btu
misc. equipment
.96 x 10^9 btu
fans/pumps
.93 x 10^9 btu
cooling
1.10 x 10^9 btu
heating
3.56 x 10^9 btu
cooling
.77 x 10^9 btu
lighting reduction
.95 x 10^9 btu
cooling reduction
.33 x 10^9 btu
heating increase
.17 x 10^9 btu

energy usage of original building

9.93×10^9 btu

lighting at

3.5 watts/ft.^{2} for 75%

1.0 watt / ft.^{2} for 25%

energy usage after 30% lighting reduction

8.82×10^9 btu

lighting at

2.45 watts/ft.^{2} for 75%

0.70 watts/ft.^{2} for 25%

note: reduction in lighting energy achieved without reducing illumination on tasks

	lighting	cooling	heating	hot water	fans/pumps	misc. equipment
	3.13×10^9 btu	1.10×10^9 btu	3.39 10^9 btu	$.42 \times 10^9$ btu	$.93 \times 10^9$ btu	$.96 \times 10^9$ btu
	2.18×10^9 btu	0.75×10^9 btu	3.56×10^9 btu	$.42 \times 10^9$ btu	$.93 \times 10^9$ btu	$.96 \times 10^9$ btu
difference	$-.95 \times 10^9$ btu	$-.33 \times 10^9$ btu	$+.17 \times 10^9$ btu	0	0	0

all btu's corrected for system efficiencies

Fig. 10. *Building energy usage reduction due to* 30% *lighting wattage reduction. Based on computer study of a building in Manchester, New Hampshire.*

replace manual operation. The new automatic control system will follow the loads much closer than it is possible to do manually; approximately another 295 MJ/m^2 yr (26 000 Btu/ft^2 yr) will be saved. The initial cost of the new control system will have a 'pay-back' time of three years.

FUTURE MEASURES:

Recommendations have been made to install full precooling coils in all HVAC units at a cost of about $70 000. The expected savings with this measure would be 1·77 TJ (140 000 cooling ton hours) per year with the corresponding reduction in energy costs. As fiscal policy permits, measures for reducing the heat gain and heat loss using additional insulation on the roof and solar control devices such as louvers, reflective coating and solar sun screen on the exterior surface of the glazing will also be implemented.

IN CONCLUSION:

In the interest of brevity I have not been able to detail all of the energy conservation measures which have been completed or contemplated. However, it is apparent that a saving of more than 50% in yearly energy consumption is readily attainable within the framework posed by management. In addition:

(1) Manpower has been held constant with no additional labour force hired because of energy conservation.
(2) Higher productivity and efficiency is being required of maintenance personnel because of energy conservation.
(3) Close evaluation of individual maintenance manpower productivity; capability and job enrichment is maintained to ensure the highest efficiency of energy conservation operation.

How does the CG Management feel about the programme? 'We are delighted with the progress, so far,' said Marvin S. Loewith, CG Senior Vice President. 'The best part about it is that these energy reductions have been achieved without any interruptions of our business and with a minimum of discomfort and inconvenience for our employees. They have been most cooperative and understanding and they appreciate that what we are doing is in the public's interest.'

Administration and Research Building for the Cary Arboretum of the New York Botanical Gardens

This is a 2-storey 2600 m^2 (28 000 ft^2) building now under construction in Millbrook, New York. It will be used as the administrative offices and research laboratories to support the arboretum programme and provide class rooms for educational purposes. The Cary Trust Fund, who donated the land and funds for the building, made some initial and far-sighted stipulations: (1) Do not disturb the land, the plant, and animal life; (2) Do not disturb the topography and aquifer; (3) Do not pollute the atmosphere; (4) Do not deplete natural resources; (5) Do not create and scatter wastes. Actually it is impossible to build any building and conform to the letter of all stipulations—but the charge was clear—minimise all adverse effects.

DMBA was commissioned to study all of the problems and implications of the trustees' admonitions. We considered all of the energy conservation measures which were previously studied for the GSA building and additional conservation opportunities such as siting structure which could not be considered for the Manchester building because of site constraints. In addition, opportunities to minimise water consumption, recycle sewage and wastes, laboratory operations, and solar energy as a major source of energy for space heating, domestic hot water heating and cooling, were explored and quantified. Working with architect Malcolm Wells, we wrote a programme for the building to meet its functional needs and the special stipulations of the Trust Fund and then designed the complete facility in accordance with those principles.

The Major Features Include the Following

The lowest level is underground and is served with light shafts to supplement electric lighting; the building is set back into a wooded area where the trees provide a windbreak on the north and sun shading on the west; the east facade is bermed up to 1·22 m (4 ft) above the ground level; masonry walls with insulation located on the external surface of the outside wall provide thermal mass and thermal resistance; the window area on the northern facade is less than on the west and east facades, and considerably less than the south facade; all

key to window infiltration chart			
(leakage between sash & frame)			
type	material	weatherstripped?	fit
① all hinged	wood-metal	yes yes	avg. avg.
② all hinged dbl. hung	wood metal steel	no no no	avg. avg. avg.
③ all dbl. hung	wood steel	yes yes	loose avg.
④ casement	steel	no	avg.
⑤ all hinged	wood	no	loose

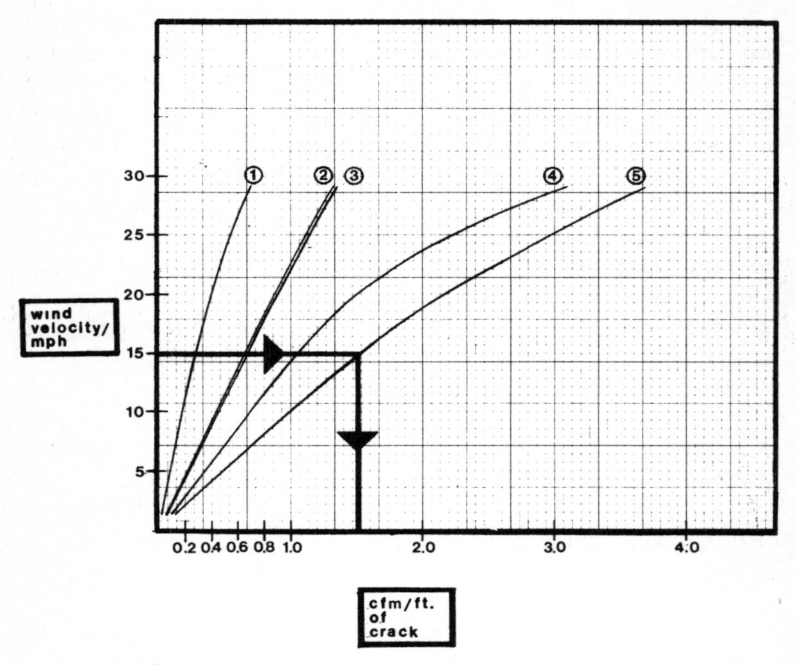

Fig. 11. *Infiltration: rate of infiltration through window frames.*

glazing is equipped with external solar control to permit the solar rays to enter the building in the winter, especially on the south facade to reduce heating loads, and shade all windows during the summer time to control the heat gain; all windows are equipped with sliding insulating shutters on the interior to reduce heat loss at night and during unoccupied periods; the U factors of the opaque walls and roofs are less than $0{\cdot}34\,W/m^2\,°C$ ($0{\cdot}06\,Btu/h\,ft^2\,°C$); the roof is designed in a saw-tooth fashion to permit the installation of solar collectors on the south facing slope of each peak; the building will be equipped with approximately 557 m^2 (6000 ft^2) of solar collectors, 3 tanks for storage of solar heat and excess heat from internal sources for use during sunless periods; a heat pump installation, using solar heat gain as a heat source supplemented with well water from a deep well, provides winter heating with variable volume zoned air heating systems; well water will be used directly in cooling coils for air conditioning and for condensing for the small amount of refrigeration needed, an experimental greenhouse will be fitted with thermal shades that can be closed at night to reduce heat loss and will be fitted with a combination water-air solar collector for heating; the same collectors will be used to temper makeup air for laboratory hoods; laboratories and all cold and hot water taps (faucets) will be fitted with flow restrictors to minimise water usage; an underground cistern will collect rain water which will be used for flushing water closets and laboratory requirements.

These and other features indicate that this building will use well under 450 MJ/m^2 yr (40 000 Btu/ft^2 yr) for all services. The solar energy system is expected to provide virtually all of the domestic hot water and service hot water requirements and more than 70 % of the yearly heating requirements. It is interesting to note that the construction budget has not been increased due to energy conservation measures, except for the cost of the solar energy system which will add approximately $100 000 to the initial cost.

Management Center, Argonne National Laboratories, Argonne, Illinois

Encouraged by the results of the GSA Manchester, New Hampshire Building, Management at Argonne Laboratories retained DMBA and Isaak & Isaak, architects, to study the building programme, conduct energy conservation studies and analyses and provide a conceptual design for an energy conserving building with the estimated energy and cost requirements for a management centre building to house 1000 people. The proposed centre is to replace existing facilities which require an expenditure of $226 000 per year in utility costs.

Using analytical methods previously developed, and the NBSLD computer load programme, the team developed a plan incorporating similar energy conservation measures employed in the GSA building but modified for the climatic conditions and special requirements for the management centre. DMBA's calculations indicate that the building can be constructed and operated to consume less than 625 MJ/m^2 yr (55 000 Btu/ft^2 yr) and save about $200 000 per year in energy costs (based on 1974 costs) as compared to the building which the new centre will replace.

A special feature of the building is a treatment of a gentle sloping south wall (Fig. 14) to be used as a solar collector for heating and cooling. By replacing the wall with a solar collector, the normal additional cost of the solar collector will be reduced by about $43/$m^2$ ($4.00 per ft^2). Storage tanks for hot water and storage of chilled water, absorption refrigeration systems operated with solar heat and double bundle condenser operating as heat pumps contribute to the dramatic reduction in energy consumption expected for this building.

Subsequent to the completion of the conceptual design, DMBA was commissioned to perform a similar study for Argonne National Laboratory for a proposed environmental evaluation laboratory. The study, completed in May 1975, indicates that measures similarly employed for the Management Center, and with special auxiliary air supply hoods for the laboratories, will produce the same energy conservation results expected for the Management Center. For the environmental laboratories, solar collectors will be located on the roof of the structure, on a south facia, and on the structure over the parking lot which serves as a car port. For both buildings, heat recovery from exhaust air, hot water drains, condenser water heat, and internal heat gain from lights and equipment, and task lighting are major contributors to energy conservation.

3*M Research and Development Center, Oakdale/Lake Elmo Site, St. Paul, Minnesota*

The Minnesota Mining and Manufacturing Corporation Management retained the Architectural Alliance team of Minneapolis, Minnesota with DMBA and Michaud, Cooley, Hallberg, Erickson and Associates, Inc., Engineers and Planners, as well as other team disciplines to develop a long range master plan for research and development facilities to be erected on a new 2·3 km^2 (570 acre) site.

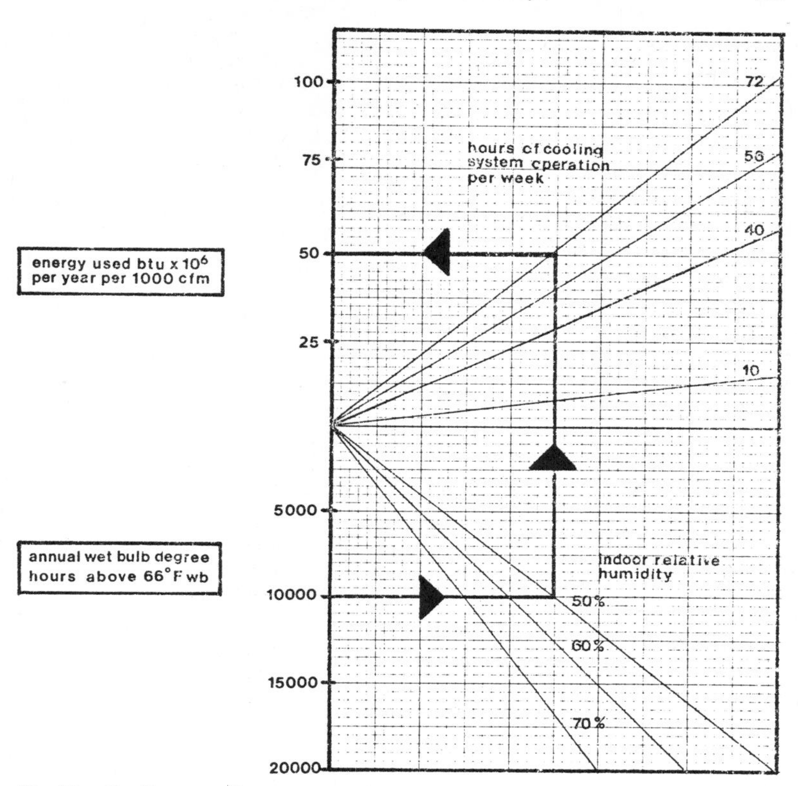

Fig. 12. *Cooling: yearly energy used per* 1000 *cfm to maintain various humidity conditions.*

The new centre contemplates 975 000 m^2 (10·5 × 10^6 ft^2) of building for research activities to be constructed in addition to their present research and development facilities which now comprise a total of 418 000 m^2 (4·5 × 10^6 ft^2).

A major requirement of the planning team included energy conservation and minimum use of natural resources. Drawing on previous experiences of energy conservation studies for candidate systems and materials, and utilising a computer programme tailored for the St. Paul climate, the engineering planning team analysed many alternatives and provided input to the architectural site and planners for building configuration, construction materials, glazing types, building height, and mechanical and electrical systems which would result in maximum energy use of 680 MJ/m^2 yr (60 000 Btu/ft^2 yr) for office space (hours of operation would be longer than for the GSA building) and $1480/m^2$ yr (130 000 Btu/ft^2 yr) for laboratories. The higher figure for laboratories is accounted for by industrial processes and the large amount of air required for exhaust hoods. The energy budgets are about 50 % of the energy used in the existing 3M Center, even though many of the existing buildings have been upgraded and fitted for energy conservation already.

Some of the salient features which are now included in the long range development plan following the studies include: 5 storey buildings, with 2 storeys underground; masonry walls with insulation on the exterior surface and a U factor below 0·57 $W/m^2\,°C$ (0·10 $Btu/h\,ft^2\,°F$), roof construction similar to opaque walls, but reflective for reduced heat gains in the summer; combined U factor for opaque wall and glazing not to exceed 0·97 $W/m^2\,°C$ (0·17 $Btu/h\,ft^2\,°F$); HVAC systems to employ variable volume boxes and controls; central refrigeration and boiler plant to serve all buildings; modular steam boilers with combustion air and oil preheated; combined turbine-driven centrifugal refrigeration units, 'piggy back' with absorption units to produce one ton of refrigeration per 4 kg (9 lb) of steam; considerations for burning oil, coal, gas and solid waste; illumination to be accomplished with an average of 22 W/m^2 (2 W/ft^2) with selective lighting systems serving each specific task; direct air supply to exhaust hoods to reduce room heating and cooling loads; heat recovery from exhaust air, process hot water, industrial heat processes, and after-coolers of large air compressors; insulating thermal barriers to shut-off windows at night during the heating season; external solar control devices combined with heat reflective glass to reduce summer solar heat gains; solar energy to be considered for domestic hot water heating, process heat and preheating make-up air for exhaust hoods.

The master plan provides a design vocabulary and guidelines for engineers, architects or in-house personnel selected to do the working drawings for the buildings. Significantly, interfacing the energy requirements for industrial processes with the environmental building requirements produces major savings in energy for both uses. Equally important, the study has shown that a selective load-shedding programme to produce peak demands can result in cost savings of more than one million dollars per year (at the present costs of electricity), in addition to the cost savings in resources and operating costs from reduced consumption.

The US Home Corporation 'Resource Saving House', Clearwater, Florida

This project demonstrates the value of energy conservation studies and the feasibility of saving energy in residential structures, as well as the larger commercial and industrial buildings. The specific goal of the owners was to build a demonstration house showing that it was possible to save 35 % of the energy and water normally consumed in a 241 m^2 (2600 ft^2) residence in Florida using existing technology and construction know-how and available hardware and appliances. A further requirement was to maintain quality and comfort, and not increase the initial cost by more that 10 % of a 'standard house' of the same size and quality.

DMBA surveyed energy usage in existing houses of the same size and type on an adjacent site in Clearwater and found that the average house consumes about 140 GJ/yr (39 000 kWh/yr) for all services, and about 1670 m^3 (59 000 ft^3) of water per year. Working with Walter Richardson AIA. of Santa Monica, California, DMBA analysed a host of proposed modifications to existing designs ranging from insulation thickness, type and location; exterior wall and roof materials; house configurations; heat pumps; lighting systems; and household appliances; we determined the potential reduction in savings by improving the performance of each of these components. The final plan for the 'resource saving house' includes the following; external solar heat gain control for all windows; masonry walls with urethane insulation installed on the exterior surfaces; vaulted ceilings to eliminate heat accumulation in attics; reflective roof surface to reduce heat gain; french doors with 50 % glazing and automatic door closers instead of sliding glass doors leading to outside patios; water-to-water heat pumps utilising ground water as a heat source; a 3-zoned heating system to permit reduced operation of the mechanical system to meet occupancy needs when they occur; low voltage lighting system control; fluorescent lighting in place of incandescent lighting; a hot gas heat exchanger in the heat pump system to heat all dometic hot water; energy conservation appliances including refrigerator, freezer, clothes washers and dryer, range and oven which will consume about 25 % less energy than the equipment most commonly used now; flow restrictors in cold and hot water lines and taps (faucets) with 0·03 litre/s (0·5 US gal/min) flow instead of 0·13 litre/s (2 US gal/min); spray type shower heads to reduce water flow without reducing the

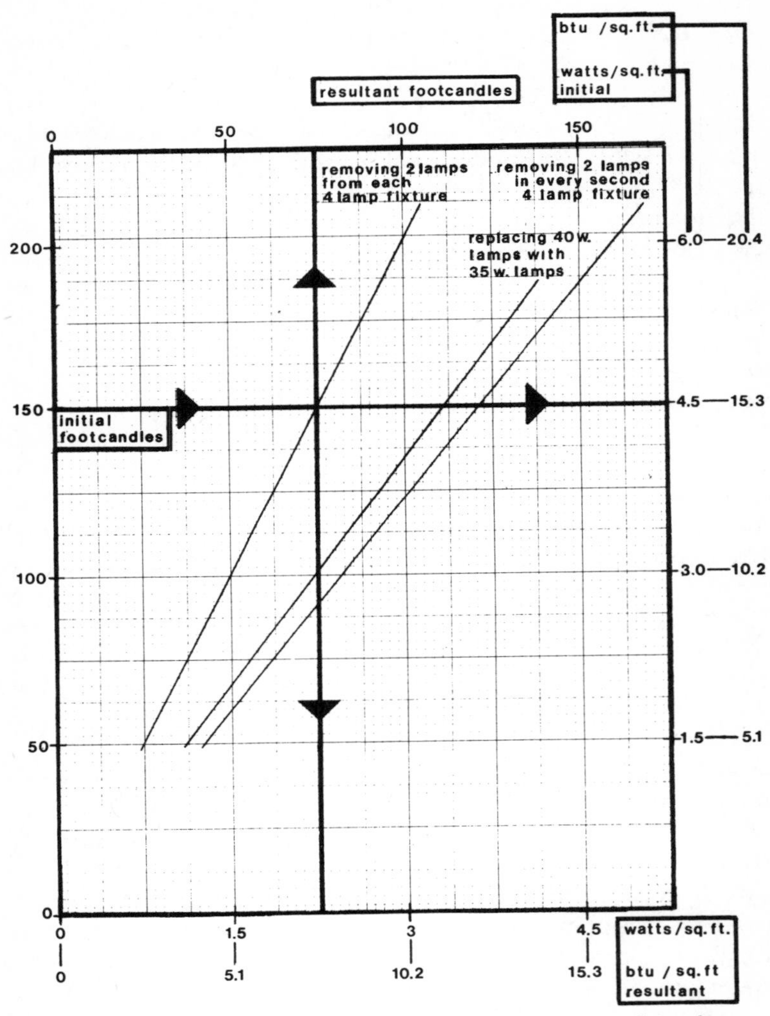

Fig. 13. *Lighting: resultant foot candles and watts/sq ft for various lamp changes.*

'comfort' of an adequate shower; underground cisterns to collect rain water from the roof for use in lawn irrigation in hot Florida climate.

Calculations indicate that the Resource Saving House will consume 86·4 GJ (24 000 kWh/yr) a 39 % reduction and 350 m^3 (12 500 ft^3) of water per year, a reduction of 78 %. The heat gain in the summer is reduced from 14—10·5 kW (48 000—36 000 Btu/h) and the design heat loss from 11·7—9·7 kW (40 000—33 000 Btu/h)* reducing the peak demand and electricity cost and permitting the use of smaller heat pumps. Management plans are to inaugurate similar resource saving demonstrations in three other climatic zones in the country.

A Forecast of Power Requirements and Reduction in Peak Demands and Consumption for Long Island, New York

This project is representative of new opportunities for consulting engineers to concentrate their expertise and time on energy related matters which will have a significant effect upon electric utility plant operations, requirements, and costs.

Under contract to the County of Suffolk, N.Y. Department of Environmental Control, DMBA first analysed the Long Island Lighting Company's (LILCO) forecast for future electrical energy requirements, demographic projects, and current energy use in Long Island, a region of approximately 2 600 000 people, and developed a detailed method of analysis for forecasting future usage which is more accurate than depending only upon historical growth. After adjusting LILCO's projections in accordance with the developed forecast methods, DMBA then selected a large number of energy conservation options which could be applied to existing and projected future buildings, and analysed them in order to determine the effect such programmes would have upon peak electric demands and yearly consumption. By virtue of having performed previous energy conservation studies, we were able to analyse the effects of energy conservation for this entire region within a three month period.

In the next phase of the study, we analysed the potential reduction of peak demand and consumption that a solar energy programme could accomplish, in addition to the reductions due to an energy conservation programme. We determined that energy consumption and peak demands could be reduced by more than 40 % below the 1995 forecast of LILCO. Seeking further ways of reducing consumption and peak demands, we considered the use of total energy systems for all installations in Long Island for buildings and developments with peak loads greater than 500 kW. These

* The standard practice house already included many energy conserving construction features of the building envelope.

installations could reduce peak demand and consumption to an even greater extent.

In the next phase of this project we worked with Professor William Heronemus of the University of Massachusetts to analyse the potential for using wind energy on Long Island. Meteorological data provided the wind patterns and velocity to determine the potential horse power available from winds on a seasonal basis.

Fig. 14. *Management Center, Illinois.*

A number of wind generator sizes and configurations were selected for analysis with two major modes of operation. (1) Using wind energy as a solar furnace to use the electricity for heating hot water to replace all electrical resistance heating and (2) to generate electricity for end use. Both on-shore and off-shore wind generating stations were considered. The results were interesting. There is sufficient wind power available on Long Island which can be harnessed with conventional wind generators, although all sizes are not generally in use today, to supply all of Long Island's electrical needs.

Detailed economic studies were not prepared. However, with escalating costs of conventional central power plants, and the restrictions on their construction due to technological and environmental constraints, it appears that generating electricity with wind for this region could be economically feasible. Additional studies have been recommended to determine the economic feasibility for such wind systems. Other measures, such as time-of-day metering, long term energy storage systems, major chilled water storage systems to reduce summer peaks, and legislation which will assure the infusion of energy conservation equipment, as well as emerging technologies such as ocean thermal differences, can assume an important role in solving the energy problems which affect the economic and social issues facing societies.

Epilogue

The value of conservation studies extends far beyond the individual study performed. In the interests of time and the limited human resources that are available, let us carefully select the conservation and research programmes which have the broadest generic value. To that end it is important that we institute a co-operative programme among the family of nations to divide this huge task into manageable parts.

The energy conservation studies and programmes outlined in this paper are by no means a comprehensive list of all of the efforts now under way in the United States, but are representative of those which we have engaged in and have found of value. Our own experiences are paralleled by many others.

While we have made some progress in energy conservation in the United States, we still have a long way to go. With the proper motivation, public education and financial support from the federal and local governments, and interest and participation by commercial, industrial, and municipal entities, we can achieve our energy goals before we are faced with insurmountable obstacles which will more adversely affect our welfare and existence.

REFERENCES

1. *GSA Energy Conservation Guidelines for New Office Buildings.* Regional Director of Business Affairs, GSA, Washington.
2. *Energy Conservation for Existing Buildings: Opportunities and Guidelines.* Vol. 1 and 2. US Federal Energy Administration, Washington.

Discussion

H. A. Rudgard (*Shell International Petroleum Co. Ltd.*): I wonder what experience is available on computer applications in buildings to conserve energy but still maintain conditions at say 20 °C (68 °F) DB in Winter and 22 °C (72 °F) DB in Summer.

J. F. Barnaby: At Connecticut General we now have on line a computer based centralised control system, initially we have saved 30 per cent of our 1972 base year adjusted, the savings are basically increase of efficiency and quick response time.

J. Keable (*Triad Architects*): I would be particularly keen to hear Mr. Dubin's views on the subject of thermal storage with heat pumps. He mentioned interseasonal storage which clearly is perhaps a necessary and practical solution in certain cases but the impression I have from some of the work we are now doing is that at the detailed level of the application of heat pumps in buildings it will be a question of just how short a period we really need for the efficient operation of heat pumps. Certainly in this country where our climate is not so extreme as in the USA this would seem to be the case.

F. S. Dubin: There are various modes and operations of heat pumps. We found, for instance, in the Manchester, New Hampshire building, as well as other commercial buildings, that to capture the energy from the core of the building—lights, people and business machines, and transfer heat to the perimeter of the building, makes a lot of sense, unless there is such a good thermal perimeter wall that heat is hardly required at all. In such a case we would suggest, and have designed, installations where the extracted heat from the core is transferred into storage in hot water tanks; the heat stored capacity can be used for night heating or morning pick up etc. The selection of a heat pump system or other types of system and the structure are integrally married together. The way the two systems, mechanical and structural, are designed together will determine the effectiveness of the heat pump system itself. Also on the chilled water side, when the heat pump is used simply as a refrigeration machine, we have found that it can be very effective to store chilled water at night time—by operating the refrigeration system at low condensing temperatures at night, stored chilled water is then drawn off during the day-time especially during the hours of peak electric load effectively to reduce the peak electrical demand.

The use of heat pumps with solar energy can be effectively done in at least two major ways—one in air-to-air heat pumps, with the solar system providing an air heat source for the heat pump, or as a water-to-water or water-to-air heat pump with solar energy supplementing other sources of energy as a heat source. In the Cary Arboretum Project in New York State we are using solar energy in combination with water from deep wells, the deep wells serving as a heat source for the heat pumps when there is no sunshine nor heat from the sun in storage for use as a heat source for the heat pump.

P. J. Bray (Denco Refrigeration Ltd): It should not be thought that in the United Kingdom we lack the ability or imagination to design and apply heat pumps and to utilise waste heat. Our company has over a period of many years designed and installed a wide variety of such plant. Several banks, offices and private dwellings have installations. It is our practice to offer heat recovery equipment with tenders for air conditioning or refrigeration when the cooling plant will be required to operate during the winter. A 2·2 kW refrigeration plant used for air conditioning will exhaust to atmosphere in excess of 10 kW of heat. The cost of installing suitable equipment within an office to make use of this heat can be recovered within two years. In the case of larger plants often in less than one year. My own office is fitted with a heat pump which makes use of otherwise wasted heat at high level in a local production area. The Coefficient of Performance of such a plant will be greater than 5—that is 5 kW of heat for each 1 kW of electricity consumed.

Our problem is that due to the current state of industry the extra expenditure for energy saving is rarely popular. I feel the Government can do much to help without incurring large costs by using its publicity machinery. If the will to save energy is sufficiently strong, many schemes will be installed which will in two or three years save the country considerable energy and create lower operating cost to industry. Surely it is more desirable to create an understanding and a will for energy conservation now rather than having to resort to restrictive legislation or offers of vast sums of money in the form of grants—later.

J. R. Parker (E. W. Herrington & Partners): I wonder about the extent of the possible extension of the life of oil supplies beyond the 25–30 years currently predicted; if the energy saving measures discussed in this conference were implemented I would have thought that if we did and were able to apply all these savings, it would be mitigated by the increased use of energy simply due to rising population. I would suggest that the contribution we could make is very marginal in relation to the total problem.

J. F. Barnaby: What we are doing is buying time. In the United States at least the technology is not coming fast enough to compensate for the amount of energy we are using.

F. S. Dubin: There are many alternatives we have not considered at this conference, for example, using ocean-thermal differences to generate electricity and in turn produce hydrogen, and to use the hydrogen in fuel cells for power generation. Wind generators can also be used to generate hydrogen. We can use the waste heat from our power plants to create our own thermal differences. We have the option of using fuel cells with heat recovery from the fuel cells with the combined cycle efficiency as high as 70% or 80%.

I am a very firm believer in solar energy. Three years ago we had never done a solar energy installation, today we have twenty designs and some installations. The potential for solar energy to handle a large percentage of our energy needs by the year 2000 is very great. Solar with conservation, I think, can see us through, and it seems to be a tremendously simpler technology than nuclear fusion. There have been improvements in photo voltaics and collectors. Within the last year a new flat plate evacuated collector has been developed. We used to say there would be an oil saving of 40 litre/yr per m^2 of collector, now with an evacuated collector, and there are several (a tubular collector has a tube within a tube, a selective surface with high absorptivity, low emissivity, and a vacuum to cut conduction and convection losses) we are looking now at an oil saving of 81 litres/year for each m^2 of collector (2 U.S. gal/ft^2 year). At present costs the collector is \$220/m^2 (\$20/ft^2). The projected costs will come down to about \$54/m^2 (\$5/ft^2) when production reaches about 150 000 m^2/annum. Because a tubular collector is like a fluorescent tube it lends itself to mass production. These types of development are tremendously heartening to come on-line about the time fossil fuels may be gone. I do not think conservation is only a short term measure at all because at any rate there will be capital costs attached to these other alternative technologies so conservation is still necessary to reduce the size and capital costs of the alternative systems.

P. R. Achenbach (*National Bureau of Standards*): In the U.S. there appear to be at least three large sources that might be available in significant amounts in twenty years or so, namely, nuclear energy, coal gasification and solar energy. Thus an effort toward energy conservation is probably most beneficial in the next decade or so. One of our national goals is to cut the annual rate of increase in energy use in half. I think we could hold the line on energy use with no increase for the next ten years by agressively attacking energy conservation. We probably will not do so well, but I believe it is possible.

8

Economic Aspects of Energy Conservation

D. FISK and S. J. LEACH

Synopsis

This paper first of all briefly considers the importance of buildings in the context of the total United Kingdom consumption of primary energy, secondly analyses the economic aspects of energy conservation in general and finally applies the analysis to two very different examples of possible conservation measures in houses namely thermal insulation and the use of heat pumps.

Energy Conservation in Buildings

Throughout the paper the unit of energy employed is the joule (J) and its various multiples MJ = 10^6 J, GJ = 10^9 J. 1 GJ is approximately equal to 9·5 therms or 278 kW hours.

The energy data used is that relating to the United Kingdom for 1972.[1] This is used instead of the now available data for 1973 because it is by and large more representative of the recent past prior to distortion from increases of energy prices and fuel shortages. The losses of energy incurred in the processing of the primary energy and its distribution are taken into account to obtain the gross energy. These energy losses are termed the energy overheads.

In 1972, the UK consumed 8·83 × 10^9 GJ of primary energy; this came from 4 sources and the amount and relative contribution of each is shown in Table 1, on the left hand side. The users of the energy are shown on the right-hand side.

Twenty-nine per cent of the energy was consumed in housing. This is nearly twice as much as the energy consumed in transport. Fourteen per cent is classified as 'other users' and this is mostly in buildings since it includes commerce and public authorities. Some of the

TABLE 1
UK ENERGY SUPPLY AND CONSUMPTION IN 1972 (UNITS: 10^9 GJ)

Primary energy supply			Gross energy consumption		
Oil	4·23	(48%)	Industry	3·60	(41%)
Coal	3·22	(37%)	Domestic	2·55	(29%)
Natural gas	1·07	(12%)	Transport	1·41	(16%)
Nuclear and hydro power	0·31	(3%)	Other users	1·27	(14%)
Total	8·83	(100%)		8·83	(100%)

consumption by 'industry' relates to the provision for the environmental requirements for workers in offices and factories and also relates to use by buildings. It can therefore be concluded that at least 40% and probably as much as half of all the primary energy consumed in the UK is consumed by buildings.

Energy Overheads

An analysis of the overheads associated with the delivery of energy to the user is described in a BRE publication.[2] The energy overheads are summarised in Table 2 where it can be seen that in 1972 the energy consumed by the oil, coal and gas industries in delivering their fuel to the users was small.

TABLE 2
ENERGY OVERHEADS 1972

Oil	7·5%	(refining)
Coal	2%	(extraction)
Natural gas	6%	(losses)
Electricity	73%	(conversion)
Substitute natural gas	26%	(estimated)

In the case of gas this relates to the overheads associated with natural gas but some gas is manufactured from oil (SNG) and this has an overhead of 26%. For the electrical industry the overhead is much higher amounting to 73% in 1972.

In considering energy conservation these overheads must be taken into account since in some cases it is possible to consume more energy at the point of use and yet save primary energy *e.g.* in the replacement of electric space heating by direct fossil fuelled heating.

Economics

In the broadest use of the term any action on energy conservation involves an economic decision. That is any conservation measure implies the consumption of some resources in order to reduce the consumption of some other and if this action is a worthwhile undertaking then the 'cost' of the consumed resources must be less than the 'cost' of the resource conserved. This paper is concerned with using resource cost as the cost basis for decision and some of its implications and problems. In doing so, it collects together some of the points raised by an earlier cost-effectiveness study.[2] A resource cost approach has the advantage over some other energy conservation appraisals, of relating directly to other economic decisions. The standard cost-effectiveness analysis contains three elements: the cost of the resource saved, the cost of the resources consumed, and some procedure for bringing costs incurred at different times to a single base. Each of these points will be dealt with in turn.

The need to resort to resource costs for energy results from the fact that a number of other resources are also finite or scarce. There may be other economic inputs such as land or labour, or just as important, social or value criteria such as the protection of the environment. It is, of course, not necessarily true that a market economy will produce a price which truly represents the resource cost. Left to itself it may, for example, shed some of the costs as 'external diseconomies' which do not appear in the price. Very often it is the social or value criteria which suffer in this respect. However an approximate balance between all resources can in principle be struck. Even if all other resources were free, however, costs would still be needed for different forms of energy. This is demonstrated by the 'rejected heat' of power generation. This heat is usually rejected at about 30 °C which is useless for most heating purposes, despite the fact that there are some 2×10^9 GJ of it annually. If, however, the rejection temperature could be raised to about 100 °C, it would become suitable for space heating, although it would be associated with a drop in efficiency of electricity generation (and hence a 'cost'). If it was raised still further, say to 400 °C (with a further decrease in generation efficiency) it would be 'too valuable' for space heating, since work could be extracted and still satisfy the space heating load. This example demonstrates that it is not necessarily energy which is in short supply but energy sources with low entropy, that is available energy. The resources to be sacrificed to save one joule of energy are thus indeterminate until the quality, low grade or high grade, of the unit of energy is known. This point of view adds further weight to the point made by Milbank,[3] that in assessing the energy consumption of a building, it is essential to differentiate clearly between electricity and other fuels. Electricity is, thermodynamically, the highest grade energy source available and therefore 1 Joule of electricity is more valuable than 1 Joule from fossil fuel. In fact the primary energy equivalent analysis used by Milbank and others represents this difference fairly well because the available energy of each of the common primary fossil fuels is roughly the same. Returning to the resource cost of fuels, it is interesting to note that the available energy plays a role here as well. If the resource cost of (net) energy delivered is reduced to the resource cost per unit (gross) primary energy required, this gross cost varies only a little between different energy sources compared with the wide variation found in the original net costs. Energy would thus seem to be delivered at roughly the same cost per unit available energy. This will be discussed later.

Having established a case for the resource costs of energy both on the grounds of competition with other resources and competition between different forms of energy, it still remains to be shown that resource costs show any sense of impending scarcity. It is not intended to discuss here whether reserves ever actually exhaust themselves, but rather to investigate whether the resource cost signals in advance any impending shortage. Clearly if the resource cost of oil at the Middle East well head was 16 cents a barrel until the day the last well went dry, whereupon it became infinite, it would be a poor indicator for the allocation of resources, particularly since energy consumers inevitably have slow response times to sudden changes in fuels in most cases. The theory of the allocation of finite resources has been well established since the 1930s and recently reviewed.[4] Not surprisingly a complete, perfectly competitive market, allocates the resources fairly well. The essential part of this idealised system is a fully developed 'futures' market in the finite resource. What 'futures' markets that do exist in the real world are more limited and it is natural to ask whether any such anticipation is included in existing resource costs.

The most useful resource cost for the energy supply industries is the long run resource cost,[5] and this is, in principle, the cost charged to the consumer by the electricity industry.[6] Suppose that a given consumer decreases his consumption by 1 W. Then the previously optimised cash flow, involving decisions on construction and commissioning of new generation plant will have to be re-optimised. In particular the introduction of a substitute energy stream will have to be brought forward, if the present energy source is finite. The difference in present value between the two cash flows, using a prescribed discount rate, is the opportunity cost of the 1 W less of consumption. The difference in opportunity costs between dropping 1 W of demand now and dropping 1 W of demand one second later is the long run marginal cost of 1 J. In more physical terms the long run marginal cost tells the user whether 1 W of extra consumption consumes less resources by expanding the supply to meet it both now and in the future or by improving the efficiency of the end use appliance. It also indicates the (discounted) saving from the delay in bringing in new power stations resulting from some energy conservation measure.

In summary, the use of long run marginal resource costs for energy do allow for competition with other resources and do indicate the extra cost implied by exhausting a finite resource. One word of caution is perhaps in order. Energy supply systems tend to show economies of scale. This means that the use of a marginal cost as a basis of decision only indicates a local optimum in the allocation of resources and not necessarily a global optimum. Thus the ambitious planner contemplating a non-marginal change might find that the total resource cost was other than indicated by the economics of a similar but marginal change. An example might be the difference between a national combined heat and electricity generation scheme, where all the electrical space heating load was lost by the electricity system and a single scheme where only one station was modified in an otherwise unchanged merit order.

The above discussion does not remove the need for energy analysis. It is clear that a resource cost cannot be arrived at, unless the full consequences of changes at the margin are known. Energy analysis is perhaps one of the most highly developed techniques in this area. In the idealised economic model energy accounting provides the crucial information upon which the market bases its values.

COST OF OTHER RESOURCES

Little more can be said, concerning the use of resource costs in this area, which has not been said in the previous section. It should be recognised that the cost of energy appears both in the energy saved and in the cost of the resources consumed. Thus a conservation measure which is just not cost-effective now is not necessarily cost-effective if the cost of energy is doubled. The savings certainly double, but so also does the cost of any energy consumed in producing the other resources. Thus the remark that 'energy resources will never be exhausted, because as costs rise, more resources will become available' is not generally true. It requires, a priori, that a technology can be developed to permit an overall available energy gain.

This argument also leads to the conclusion that traditional cost-effectiveness analysis covers the possibility of a measure consuming more energy than it saves. In general terms cost-effectiveness implies that: cost of fuel × fuel saved > cost of resources consumed × discount factor. The important thing about the discount factor is that it will be less than unity. If the cost of energy in all processes associated with the consumed resources is separated out then this inequality implies: cost of fuel × fuel saved > cost of fuel × fuel consumed. If the cost of fuel to the producer is equal to or larger than the cost to the consumer the fuel saved always exceeds the total fuel consumed. If the fuel was cheaper then clearly the manufacturers would be better off selling fuel. However fuel of too low a grade, or environmentally unsatisfactory for use by the consumer, are both reasons for breaking the inequality. It is important that the costs are historically equivalent. Even between fuel industries the response times to a given price rise in one commodity may vary greatly.

DISCOUNT RATE

Apart from periods of price restraint and local market distortions, resource costs and prices do not diverge from each other markedly. More fundamental distinctions arise when considering the discount rate to be used. In the public sector the discount rate is set at 10% in real terms.[7] This is intended to represent the opportunity cost (after allowance for taxation differences) of capital in the private sector. In the strictest terms the discount rate represents the return on the next investment that could be made (the marginal investment) if the conservation measure was not undertaken. For firms with good management accounting a corporate discount rate is thus likely to exist. The most difficult rate to determine is that of the individual consumer, because the opportunity cost varies, depending for example on whether he raises a loan or uses cash. The alternative marginal investment may not necessarily be a deposit account or building society account and indeed this will not necessarily produce the best return in real terms.

The important point about a correct discount rate is that it represents the opportunity cost of not undertaking another policy or action which has a defined rate of return. The existence of this alternative makes possible the deduction of conditions for investment that reduce the need to speculate in detail concerning the future cost of energy.

TIMING OF ACTION ON CONSERVATION

If it is assumed that the cost of energy in real terms is not likely to fall then the optimum time for investment can be stated, assuming that the capital is kept in an alternative investment earning interest at the discount rate until the appropriate time for action is reached. In general the optimum strategy is to keep the capital in the alternative investment until energy costs have increased sufficiently for the conservation measure to become cost effective. Thus a measure which is not cost effective at current costs should not, on cost-effective grounds, be implemented.

ENERGY COSTS

The following figures represent the best assessment of average domestic energy costs that BRE could make in a rapidly changing situation, after appropriate consultations with, and taking account of the views of the Department of Energy. The figures are inevitably rather approximate: they are intended to reflect costs in the early part of 1975, but are assumed to prevail over a whole year. There is, however, considerable variability in the costs to supply energy to different consumer classes and sizes around the averages shown.

Gas	£1·23/GJ (net)		
Electricity	£3·98/GJ (net)	Off-peak	£2·78/GJ (net),
		On-peak	£4·17/GJ (net)
Coal	£0·95/GJ (net)		
Oil	£1·52/GJ (net)		

Using the energy overheads derived in the report[2] the cost per GJ (gross) of energy is:

Electricity	£1·05/GJ (gross)
Coal	£0·93/GJ (gross)
Oil	£1·40/GJ (gross)
Gas	£1·04/GJ (gross)

In terms of primary energy the four costs for coal, oil, gas and electricity are close and there are reasonable grounds for taking £1·05 per GJ as a cost for primary energy except when directly comparing two fuels. This cost is used in the calculations below.

EXAMPLES OF ENERGY CONSERVATION MEASURES

Two specific examples—Thermal Insulation and Heat Pumps for housing—taken from the report of the BRE Working Party on Energy Conservation in Buildings[2] will now be discussed and the general conclusions of the report summarised.

Thermal Insulation

Thermal insulation saves energy irrespective of how the space heating is supplied and is also important to thermal comfort. The Report of the BRE Working Party concluded that if the existing housing stock had been cavity filled where possible, if the loft insulation had been improved, and windows double-glazed, the UK energy consumption would have been 3–4 % less taking account of the past evidence that some of the potential fuel saving in older properties with only partial heating, would have been taken up in increased comfort. If the full potential had been realised the energy saving would have been about 5 %.

The cost effectiveness of thermal insulation of housing will now be considered from the national point of view and secondly from the viewpoint of the householder.

An improvement in thermal insulation results in a redistribution of the temperatures within the dwelling and a reduction in the cost of heating it to a required level. The estimated saving that could be achieved without any change in internal mean temperature is given in Table 3 together with the savings that could be achieved when the ground floor temperature alone is held constant.

The 'theoretical' saving is that realised for unchanged mean internal temperature. The 'controlled' saving is that realised when ground floor temperature alone is held constant by thermostat control before and after insulation. The difference between the reduction in heat loss and the reduction in fuel consumed, results from the contribution of fortuitous heat gains to the maintenance of the inside temperature. In

TABLE 3
ESTIMATED SAVINGS FROM INSULATION OF WELL-CONTROLLED DWELLINGS

	(22·8 *cm brick outer wall*) *Loft insulation*		(*Cavity walling*) *Cavity and loft insulation*	
	Theoretical saving	*Controlled saving*	*Theoretical saving*	*Controlled saving*
(i) Heat loss	15%	6%	32%	24%
Fuel saving	19%	8%	40%	33%
(ii) Heat loss	15%	9%	32%	27%
Fuel saving	19%	11%	40%	35%

(i) House heated only on ground floor
(ii) House heated on both floors

a poorly heated dwelling savings may be taken up in improved thermal comfort.

Loft and cavity insulation gives a potential saving of 40% of the original fuel consumption, and loft insulation alone in an older 22·8 cm brick walled dwelling 20%. In those dwellings without a good standard of heating and control only about a third of the potential savings are assumed to be realised. A reasonable resource cost for cavity fill and loft insulation might be about £120/dwelling, although tenders for groups of dwellings treated at the same time might be as little as £70–£80. For dwellings with poor heating standards and control the realised savings will be about 13% of the annual fuel input. This is estimated to average out at about 7·5 GJ/annum of primary energy for each dwelling. The cost-effectiveness of this measure is investigated in Table 4.

TABLE 4
COST-EFFECTIVENESS OF CAVITY FILL AND LOFT INSULATION

	Cost (present value)	*Energy savings (present value)*		
		0%	4% *pa*	7% *pa*
(a)	70	78	132	227
(b)	120	77	122	183
(c)	120	308	488	751

Cost-effectiveness of combined cavity fill and loft insulation for:
(a) New house with treatment at time of construction.
(b) Treatment of existing house with 40 years life left.
(c) As (b) but for a dwelling with a high standard of thermal comfort and heating control.

Table 4 shows that for an existing house with 60 years life left the measure is cost-effective at between current costs and a fuel cost rise of 4% per annum and for an existing house with 40 years life left just cost-effective at the 4% cost profile. This latter period embraces the great majority of the 8 million housing stock with cavity walls.

The cost-effectiveness is considerably improved in dwellings with a high standard of thermal comfort and control. The controlled savings are estimated to amount to about 30 GJ of primary energy per dwelling. The measure is now cost-effective at current costs both for existing and new dwellings.

Insulation of a new dwelling at the time of construction is likely to be considerably cheaper, perhaps less than £70/dwelling. This is cost-effective at current costs even when there is the likelihood of some savings being taken up in improved comfort. It is certainly cost-effective at the 4% fuel cost profile. This measure would have a nominal resource cost of some £25 M/annum and, assuming the same 'mix' of heating systems as prevails currently, would reduce the national consumption of primary energy by more than $4{\cdot}3 \times 10^6$ GJ each year. It is in fact likely that the heating systems actually installed will be better controlled, and certainly better sized to the new level of insulation. Thus it is not unreasonable to expect that in many cases the realised saving will be close to the full potential value.

The remainder of the housing stock—those without cavity walls—may, in general, be given loft insulation and a reasonable estimate for the cost of this operation performed by a contractor might be £45 per house. Table 3 indicates a potential saving of about 19% of the energy input, and thus a realised saving of about 6%. The cost-effectiveness of this measure is investigated in Table 5, for the case of a dwelling with 30 years useful life left.

Loft insulation is at contractor's cost, 'do-it-yourself' cost would be about £30. Both values assume some loss of savings to improved thermal comfort. It can be seen that this measure is just about cost-effective at current costs. If the work is carried out by the house owner the cost probably falls to about £30. The measure is then certainly cost-effective from the national point of view at current costs. Table 3 indicates that even with well controlled heating, the realised savings are not likely to exceed 10%.

In estimating the cost of double glazing, it will be assumed that the casements had previously been treated to reduce draughts so that energy savings associated with the double-glazing are limited to the

TABLE 5
COST-EFFECTIVENESS OF LOFT INSULATION AND DOUBLE GLAZING

	Cost (present value)	*Energy savings (present value)*		
		%	4% *pa*	7% *pa*
Loft insulation of solid walled house (over 30 years)	45	45	67	95
Double glazing (over 60 years)	250	31	53	91

reduced conduction losses alone. It will also be assumed that the major losses occur when the window is not curtained. On this basis a figure of 10% potential savings and 3·3% realised savings has been used to form Table 5. This table considers the least favourable case, of a house recently built with 60 years life left, double glazed with provision of adequate ventilation for a cost of £250. The savings of 3 GJ/annum do not make the measure cost-effective even with the severest cost profile. It is only in the most favourable case, that of a new dwelling heated to a high standard and with good controls that the measure is just about cost-effective and even then only at the severest cost profile.

This latter conclusion should not be taken to imply that an individual householder cannot make his house cheaper to heat by undertaking double glazing—only that if performed by contractors it is not likely to be cost-effective. Other benefits do accrue such as improved thermal comfort, and with proper design, improved sound insulation but these benefits have not been valued here.

To summarise, cavity fill and loft insulation are cost-effective at current costs for a dwelling heated to a good standard of comfort with good controls. For new construction the measure is likely to be cost-effective at current costs even if some of the savings are likely to be taken up in improved thermal comfort. For other dwellings with cavity walling the measure is cost-effective at above the 4% fuel cost profile. Loft insulation of the remainder of the dwellings by contractor is marginally cost-effective at current costs though certainly cost-effective at current costs if the householder undertakes the installation himself. Double glazing is not cost-effective at any of the fuel cost profiles except in the case of a well heated, well controlled dwelling where it is just about cost-effective at the severest fuel cost profiles.

The Householder

The fuel cost to the consumer is that determined from its market price. The relative pricing of fuels is influenced by a number of considerations and there may be times when the market price will not reflect the resource cost. Secondly additional gains or losses might arise. For example, an increase in rateable value on completion of cavity fill and double glazing, if incurred, might represent an increased rate burden to the owner or an increased market value to the landlord. Such additional gains and losses appear as transfers within the national analysis and are thus not counted in the national total. Thirdly the discount rate applicable to the consumer is not necessarily the Treasury Discount Rate of 10 %. The discount rate required is a measure of the opportunity cost of tying up capital in, say, a new heat pump. The usual method of determining this opportunity cost is to compare the fuel saving with the return on long-term secure investment of the same capital sum. Allowance must be made for tax, so that, in general the rate of return on an investment is not equal to 10 % in real terms. Fourthly, any fuel savings may be taken by the householder in improved thermal comfort and hence reduce the gains from the energy conservation measure. The value to the occupant of such comfort gains must approximately equal the fuel saving sacrificed to obtain them. Thus the analysis can be modified to include the benefits of improved thermal comfort by taking the fuel saving as that which would occur if the comfort standards remained unchanged before and after a measure was undertaken.

The points so far discussed refer to modifications in the general structure of the analysis of cost-effectiveness. If, however, the purpose is to determine the favourability of present market forces a number of more serious problems arise. For many consumers the deciding factor for a given measure will not be a positive total present value, but rather the size of the initial capital outlay or the repayment period on the required loan. Thus monetary measures, for example credit controls, will play a major part. Similarly the majority of credit sources available to the consumer will require a 'payback' period considerably shorter than that of the measure itself, imposing a further restraint via the domestic budget against large capital projects. These considerations will have considerably less influence in new private dwellings, since they will effectively be financed through the mortgage on the property at a lower net rate of interest over a longer period.

However, the consumer might still not take action on a measure with a zero present value because a greater value could be obtained by delaying action until the price of fuel rose further. As far as the consumer is concerned, for action in existing dwellings, the measure must recover its cost in its own lifetime. In fact, for many cases, if the capital cost is appropriately amortised as a cash flow over the life of the conservation measure, the optimum policy is that for which the first year's gains equal the first year's amortised cash flow.

The discussion so far has presupposed the existence of a suitable discount rate. During the previous decade, the return on secure investments, after tax, has only been a few per cent in advance of the rate of inflation. At present the return may be a few per cent less. Under these circumstances there are extreme difficulties in interpreting the opportunity cost of capital as the interest rate on the capital market. A different approach has therefore been adopted, which goes some way to meeting the above objections without sacrificing the objectives of this section. The methodology employed is that commonly used in consumer publications: The annual savings (at current prices) are compared with the annual interest obtainable from an investment of the capital cost in a building society (7·5 per cent, tax paid) and a depreciation charge equal to the capital cost divided evenly throughout the estimated lifetime of the conservation measure. The result of such an analysis does not, of course, determine the economic optimum although it may not be far from it. Implicitly, the opportunity cost is that for a householder with the capital to carry out the measure. The opportunity cost of capital would be higher if a loan had to be undertaken. In general the analysis provides an indication of the conclusions likely to be drawn by the well-informed householder. A calculation is also needed of the time to recover costs. If this is comparable with the expected occupancy of the dwelling, typically a total of 8–10 years, the measure may not appear attractive unless some of the cost can be passed on to the next occupier.

As before, prices refer to early 1975 and have been taken as roughly equal to the resource cost figure given earlier. The total cost to a consumer will also depend on tariff and standing charge considerations. Thus the choice of gas cooker is particularly favourable when the user is already on a gas tariff with a low 'run on' cost because of gas consumption by a central heating system. These detailed considerations will not be dealt with here, although they need, of course, to be borne in mind by the individual consumer. The consumer analysis will be performed for the case of the choice of new appliances or insulation treatment. This avoids the uncertainties in estimating the value to the consumer of an existing appliance.

It was calculated in Table 3 that cavity fill and loft insulation gave a saving of 35 % of the annual fuel consumption at constant thermostat settings. The average fuel bill for 1975 based on a primary energy cost of £1·05/GJ (gross) is just under £80. As a specific example, to meet a demand of 50 GJ of useful energy at 1975 'Gold Star' tariff would cost about £80 after deduction of standing charges. Cavity fill and loft insulation thus represent a saving of £28. Some of this saving may go towards improved thermal comfort standards, but the benefit of this improvement in comfort will have a value approximately equal to the savings sacrificed. The total gain on improvement of insulation thus remains at £28. Against this saving is to be set the interest lost on capital, which for insulation treatment costing, say, £120 is £9 per annum. The total saving is thus about £19 per annum which gives a period of about 6 years before the initial capital is recovered.

For houses heated to a lower standard a demand of 35 GJ of useful energy saves only about £20 on the 'Gold Star' tariff and takes about 10 years to repay. The saving for loft insulation with fixed thermostat setting was estimated earlier at about 10 %, a saving of £8 for a fuel bill of £80 (after deduction of standing charges). The laying of loft insulation might cost about £45 if performed by a contractor, but only about £30 if performed by the householder himself. Capital charges in the case of a contractor amount to about £3 per annum, and have a net saving of £5 per annum. The measure does not, therefore, look particularly attractive if undertaken by a contractor because of the long period before the cost is recovered. Undertaken by the householder himself however, the net saving is closer to £6 per annum and the measure then looks about as attractive as cavity fill.

Double glazing was estimated to reduce the heat loss by about 10 % or £8 per annum on an annual space heating bill of £80. The capital charges on an installation cost of, say, £250 amount to about £19 and therefore exceed the potential savings at current costs. It would take over twenty years at the medium fuel cost profile before the savings began to break even with the capital charge, and considerably longer for the capital outlay to be recovered in a reasonable length of time. This result means that there is very little incentive on the basis of fuel cost savings to install double glazing. The other benefits have been mentioned previously.

Heat Pump

The heat pump is a device which extracts thermal energy from a low temperature source and upgrades it to a higher temperature so that it then becomes useful, for instance for space or water heating in buildings. The low temperature source may be the air, the ground, a flow of water, or reject heat from an industrial plant: in fact any medium from which energy at a reasonable temperature can be extracted. The main feature of the heat pump is that it will always provide more energy for heating than is used in driving it. The principle is similar to that of a domestic refrigerator; both extract heat from a source, at an absolute temperature T_1 and give it out at a higher temperature T_2, a certain amount of work having been done on the system.

The most useful parameter for comparing the behaviour of heat pumps is the coefficient of performance (COP) which measures the ratio of the heat output to the energy used to drive the machine. It is the fact that this ratio can be several times greater than unity which makes the heat pump attractive when energy economy is sought.

The COP varies inversely with the temperature difference ($T_2 - T_1$); with a source at -1 °C and output at $+40$ °C the theoretical COP is 7·6. In practice, the value is less than this, partly because of the losses associated with the mechanical equipment of the pump, but mainly because of the finite temperature differences necessary to transfer heat at the source and sink (output) and the physical properties of the refrigerant used.

An unpublished BRE survey of practical experience with heat pumps in buildings in the UK over the post-war years has shown that current machines will operate with COP between 2 and 3 averaged over the year. It is feasible that future heat pumps may have coefficients of performance greater than 3. The report of the BRE Working Party on Energy Conservation in Buildings[2] assumed an annual COP of 3 for a heat pump supplying the domestic space heating load and concluded that the resulting annual national consumption of primary energy for this purpose could be reduced from the present $1{\cdot}4 \times 10^9$ GJ to about $0{\cdot}8 \times 10^9$ GJ. This saving from using heat pumps would be about 7 % of the total national primary energy consumption. A further 2 % would be saved if water heating were accomplished in a similar way but with a COP of 2, the reduced figure arising from the need for a higher output temperature.

These are large energy savings, but even these figures do not indicate the total potential of heat pumps, which can also be used to reclaim heat from outgoing ventilation air and waste hot water. Furthermore applications in non-domestic buildings are already well established and may be expanded considerably.

The obvious and straightforward uses for heat pumps in domestic buildings are to supply space or water heating. Because the COP of a pump depends inversely on the source/sink temperature difference, most space heating applications use warm air rather than hot-water radiators as the house-heating medium, as the output temperatures can then be lower.

Heat pumps can be made reversible so that they can also extract heat from the building and reject it outside, thus acting as air-conditioners. All the domestic heat pumps made in the USA are of this type, and indeed were developed from the package air-conditioners. Although this facility would only be sought in rare cases in the UK climate the reversible heat pump has properties which particularly suit it for coping with the heating and ventilating demands of other buildings of extended or pavilion-type plan form. A number of air-to-air machines can provide individual zone heating or cooling with mechanical ventilation.

TABLE 6

COMPARISON OF A HEAT PUMP WITH A GAS HEATING SYSTEM

	Heat supplied p.a.	*Primary energy usage p.a.*	*Cost p.a.*
Heat pump (electric COP = 3)	50 GJ	61 GJ	£64
Substitute natural gas fired boiler	50 GJ	105 GJ	£110
		saving	£46 p.a.

In more compact buildings, water ring-mains can be used either as source or sink for a number of individual room pumps so that, again, individual zone control can be achieved with the bonus that if certain parts of the building need cooling, the energy extracted may be used to heat other parts. In large office blocks where the adventitious gains (solar, lighting, from the occupants, and from equipment) are high this may give major savings. A central heating/cooling plant on the ring main is needed to cover any imbalance. In other cases a modern building may have a central plant for processing ventilation air. The exhaust air from the building carries reject heat and a heat pump is one method of reclaiming this, exchanging the heat with the incoming air.

Uses for heat pumps arise when attempts are made to produce dwellings whose energy demand is made much less than that of a conventional well-insulated house. Such dwellings have more thermal insulation, so that the adventitious heat gains and ventilation heat losses become relatively more important. Where, as is usual in such projects, attempts are made to reclaim 'waste' heat, or to use energy from wind or solar radiation, heat pumps are a common feature of the energy systems.

Natural energy sources, in climates such as the UK, need to be linked to some form of storage and the storage in turn to the space or domestic hot water heating systems. Heat pumps can form convenient links. Three designs for experimental low-energy houses to be built at BRE all make use of heat pumps. One has an air-to-air machine which supplies all space heating, and recovers heat from outgoing ventilation air. A secondary heat pump, using part of the output of the space-heater as its source, heats the water for washing. The second house featuring heat reclaim, uses heat exchangers for ventilation heat recovery but has a heat pump recovering the heat from waste hot water held in a catch tank. The third house, which has a solar collector linked to thermal storage uses no less than three heat pumps; one to upgrade heat from the solar collector into storage when radiation is low: one to upgrade heat from storage into the space-heating system: and one to upgrade heat from storage into the hot water system.

The cost of running a domestic heat pump compared to a gas fired boiler system is examined in Table 6 for an electrically driven heat pump of COP = 3 and a gas boiler operating at 60 % efficiency, using substitute natural gas. The cost of primary energy was taken to be £1·05/GJ. The energy requirement was taken to be the annual average domestic space heating load.

The heat pump installation would cost about £400 as against £100 for the gas boiler for a saving of present value £289. In the event of large scale production of heating only heat pumps, their cost might be expected to fall in real terms.

At the present time, the replacement of existing gas boiler systems by heat pumps is not cost effective unless the energy costs increase at

4% in real terms. However, the installation of heat pump systems in new housing, in existing housing receiving central heating for the first time and in replacing systems at the end of their life would be more favourable.

Large scale introduction of heat pumps would require their introduction in existing houses. One possibility is being studied at BRE together with absorption cycle machines and machines with non electric drive and boost. As has been mentioned earlier the optimum time for implementation of a particular measure is that point in the future at which it first becomes economic. However, the research and development programme for the technology concerned must anticipate this so as to be able to satisfy the market when this time is reached.

Acknowledgement

The work described forms part of the research programme of the Building Research Establishment of the Department of the Environment and this paper is produced by permission of the Director.

REFERENCES

1. *United Kingdom Energy Statistics* (1973). HMSO.
2. *Energy Conservation:* A study of energy consumption in buildings and means of saving energy in housing. Building Research Establishment Garston, CP56/75. June 1975 p. 65.
3. Milbank, N. O. (1974). Energy Consumption in Tall Office Buildings. 'Tall buildings and people' IABSE Conference Oxford.
4. Solow, R. (1974) The economics of resources or the resources of economics. *American Economic Review*, **64** (2), 1–14.
5. Turvey, R. (1969). Marginal cost. *Economic J.* 282–299.
6. 'Electricity Prices and National Resources'. Evidence to the Select Committee. Electricity Council, July 1974.
7. *National Industries.* (1967). A review of economic and financial objectives. House of Commons Cmnd 3437, HMSO.

Discussion

A. P. Josephides (*Property Services Agency*): In local and central government housing built to Parker-Morris standards (where the requirement is only for partial heating) is there any reason to believe that off-peak hot air from an electrical storage heater, properly designed-in, will be any more expensive than any other type of energy? As public housing comes under stringent Treasury cost limits, capital costs of the installation, as well as running costs, should be taken into account in any form of energy to be used. Have we any significant feed-back to demonstrate whether this type of off-peak electrical energy is, in the long run, more expensive than gas, taking into account capital, maintenance and running costs?

Having produced well-insulated, well-heated houses, how do we know that the tenant will make good and economical use of the installation? The delegates at this conference are an intelligent audience who can afford to heat their houses, but the great mass of people are frightened of escalating costs and unable to use prudently their appliances. Unless they are properly educated and made aware of the finer points of their heating installation, many of our energy-saving techniques will not materialise.

Dr. S. Leach: Work has been undertaken jointly by the Electricity Council and the Gas Corporation to directly compare Electricaire* central heating and gas warm air central heating both delivering the same amount of useful heat. This has shown, very roughly, that the net energy use of the two systems is approximately that Electricaire uses 80% of the net energy of the gas system. Going back to primary energy means that the gas system is much more efficient. This means a large difference in running costs and it is quite clear that the gas warm air system will be cheaper to run. The conclusion of our analysis is that the gas system will be significantly cheaper in the long run.

This is complicated by a point which I think may well come up in a number of other ways if one talks about the long term use of gas. Current gas prices are based on North Sea gas—this has almost certainly a shorter life than the North Sea oil. So if we are looking in the long term we ought to be looking not at North Sea gas but gas which is obtained in some other way and this will be more expensive. If we drastically simplify the problem we are talking about the alternative of using oil in our power stations at no more than 27% efficiency and using oil to produce a synthetic natural gas for which technology now exists and which has a conversion efficiency of about 70%. We come back to the conclusion I mentioned at the beginning—that the gas system would be cheaper in the long run.

E. J. Anthony (W. S. Atkins & Partners): We have talked about incentives to energy economy. We might also have discussed disincentives and the removal of disincentives. I would head a list of disincentives with Parker Morris with its minimum standards which do not in the long term result in energy saving; in that comparative assessments of heating systems based on Parker Morris yardsticks favour the use of sophisticated fuels such as gas and electricity for heating, whereas higher standards would be an incentive to use lower grade fuels and waste heat.

Another disincentive to energy conservation, in that it is also restrictive to total cost assessments, is standard cost plans for schools and public buildings based purely on precedent?

* 'Electricaire' is the Electricity Council's Trade Name for their Electric Storage Warm Air Heating Systems. Ed.

G. Haslett (Electricity Council): I would like to take up Dr. Leach on a number of points. Firstly, on the recommendation that electric resistance space-heating should be replaced by fossil fuels. Most of the domestic electric space-heating is in the form of storage radiators using off-peak electricity. Approximately 70% of these installations have between 2 and 4 heaters and an average installation of 3 radiators has an installed load of 8·5 kW. By no stretch of the imagination does this constitute full heating. If such a system was replaced by a gas fired system it would be unreasonable to expect a customer to pay for another background system to be installed. Almost certainly a full central heating system by hot water radiators combining the domestic hot water service would be installed and the customer might be expected to use it as such.

Our field trial evidence shows that a typical semi detached dwelling of 80 m^2 floor area would have a consumption of 48 GJ for storage-heating, some direct 'topping up' and water heating. The same dwelling replaced with a gas fired hot water radiator system would inevitably maintain a higher standard consuming for the same services between 130 GJ and 148 GJ for a range of seasonal boiler efficiency for space-heating of 50–60%. Primary energy consumptions would therefore be very similar and no energy would be saved. Dr Leach's basis of comparison of maintaining similar standards is therefore an inappropriate one.

Secondly, we are all agreed that good thermal insulation is a first need. However, most gas boilers are likely to reduce in efficiency in an insulated house due to the greater hours of operation on part load. Evidence of this may be obtained by analysis of results in the Building Research Establishment's current paper (CP20/74)[1] which indicated a drop in seasonal efficiency from 60% to 50% when insulation was installed.

In a super-insulated house having a gross seasonal energy requirement of 8000 kWh, at least 5000 kWh might be expected from miscellaneous gains. How would fossil fuel systems be designed to provide the remaining small heating requirement efficiently? Of course, the system efficiency of the water heating service supplied by a central boiler is very low. The South Eastern Gas Board[2] have shown it to be in the order of 20%. In fact the seasonal efficiencies quoted for fossil fuel systems in the BRE current paper CP/56/75[3] on which this paper is based are extremely optimistic and are not based on field evidence.

Finally it is stressed in the BRE current paper 56/75 that the conclusions are interim in nature and should not be taken as recommendations but only as a discussion paper. Is Dr. Leach speaking personally or have the BRE hardened in their views?

[1] Whiteside, D. *Cavity Insulation of Walls: A Case Study* (1974). Building Research Establishment, Garston, UK. CP20/74, p. 25.

[2] Emerson, J. and Roberts, J. P. *Summer Hot Water from Central Heating Boilers.* (1967). Inst. Gas Eng. Communication 757, p. 21.

[3] 'A Study of Energy Consumption in Buildings and Possible means of saving energy in housing', Building Research Establishment, Garston, UK. Paper CP56/75, June 1975, p. 64.

Dr. S. Leach: The first point was that the replacement of a partial electric heating system by a gas central heating system could provide a better service and yet would use the same amount of primary energy. This is essentially what we are saying but we have taken the comparison of providing the same service. I think Mr. Haslett is agreeing with us here that there is a more efficient use of energy through use of a gas system instead of an electric system. I do not know now whether anyone from the Gas Corporation would like to comment on the latest system they are developing—'Hot Line'—which is much closer to the replacement of individual electric storage heating and which provides the possibility of giving the same service.

The second point was that in one of the references we quoted there was a reduction in appliance efficiency following insulation and because the appliance was then running at a lower load factor and therefore a reduction in the expected savings of fossil fuel. We agree with this absolutely. We might perhaps argue a little about the magnitude of the changes in efficiency and on the absolute values of the efficiencies that are calculated but the point in principle is right. However, we come back to the fact that electricity can only be at the best 27% primary energy efficient and in the case of block storage heaters we are probably running at something less than, perhaps 19–20% primary energy efficiency. We can lose quite a lot in efficiency with a fossil fuelled appliance to get to that sort of primary energy efficiency. Whether or not it is 50% or 60% still means that there is a factor of two better for fossil fuelled appliances than in the electrical system.

In relation to the efficiency of gas water heating for washing purposes we tried to get information on this, we looked around to see what information was published and we talked to people in the industries concerned and we made what we believed to be the best guess. We are now currently carrying out research specifically to find the efficiency of heating appliances in summertime conditions by experiment.

On the point about the 'super insulated' house, I am very glad you have raised this as we have been pointing out to the various fossil fuel industries that when we do move to the higher standards of insulation there must be appliances on the market to meet the demand.

On your final point—have we produced a discussion document? BRE does not issue government policy on the spending of treasury money. We are providing part of the information on which decisions could be based.

F. L. G. Hartgroves (Department of the Environment): I am responsible for some very large boiler houses and central heating stations and of course we are in a terrible dilemma over the choice of fuel for these. In the paper there is reference to substitute natural gas. I would simply like to ask, are they in fact referring perhaps to the hydrogen economy. It seems to me that the advantages of burning hydrogen are so manifest, and it does give us some comfort to think that the installations we are now installing will be of use when the crunch comes and fossil fuels too expensive to use.

Dr. D. Fisk: The concept we were using in our study is a fuel generically known as substitute natural gas which as Dr. Leach mentioned earlier can be produced from oil. The technology already exists to obtain it from coal as well, so that in view of the existing coal resources it is close to what the Americans call a 'backstop technology'.

Mr. Hartgroves said he was very much concerned on grounds of availability as to what fuel to use—there have also been a considerable number of remarks asking for detailed fuel price predictions from the Department of Energy and others asking for an energy policy—though I am not quite sure what they mean by that in specific terms. I would like to expand the question to discuss exactly how I would suggest we should be looking at this as a design problem. Firstly, if a measure is cost effective now then there is little problem—'let's go out and do it'. The interesting and difficult cases are those that do not appear to be cost effective at existing energy costs but which you may think are going to be cost effective in the future because, for example, energy costs might rise or the availability of fuels might change, which is essentially the same problem. When we are looking at what, with due deference to the American War of Independence, we might call the 'retrofit' situation, the problem has a fairly simple solution. One example of a 'retrofit' might be the installation of double glazing when the original pane is already there. In this situation it is quite easy to show that you only need to ask for one undertaking from your Department of Energy advisor and that simply is that the cost of energy will not drop in the future. Then the optimal time to undertake the 'retrofit' is on the particular day that the energy prices have risen to a level at which the measure is cost effective even if energy costs were not to rise further. The really tricky case and the one I really wish to emphasise is where it is not, at existing prices, cost effective to undertake a conservation measure but where if you do not undertake the measure now and you return to it sometime later the cost of undertaking it will have considerably increased. These are usually the type of problems that arise at the design stage. Many measures are cheaper to install at the construction stage than as a 'retrofit'. What I am really suggesting is that you should be introducing into the cost effective analysis what economists would call an 'option cost' or in other words an 'insurance premium' and that one should be looking at the design stage at what the design options actually are. For example, if natural gas is available now and in some subtle way you design the heating system so that if it were not available at any other time you could not possibly use another fuel, then indeed you have made a rather poor design decision. You have written off a large number of options. If on the other hand you go to a small expense to keep those options open, then for that small option cost you give yourself flexibility. Option costing (if I can adopt the term that the French economists use) can give us a number of clues. I would like to give four examples.

Firstly, if costs are equally balanced between better insulation of a new dwelling or going for a more technically advanced appliance, then invariably option costs lean towards the insulation of the building. That is simply because the appliance is going to be replaced after say 10 years, but in ten years time when you came back to try and improve the U value of the wall you would find that you have to incur an enormous expense.

As a second example, if you are in a position where using individual appliances in a large open development is economically balanced against using district heating mains then the latter have the option cost advantage because they are capable of using a vast range of fuels. It is obviously much cheaper to change fuels in the future at a central boiler house than at each of the individual house boilers in an estate. Indeed you could realise option cost in sheer money terms by purchasing for the boiler house a multi-fuel boiler.

Thirdly, if you are going to choose the way in which the heat is supplied to the house, then the lower the delivery temperature of the heat supply, the bigger the number of options you can accrue for yourself for the future. If you are a 'priest' of solar energy systems then avoid building houses which need extremely hot water in the radiators because this will write off a large part of the solar energy retrofit design option.

I think insulation of dwellings is perhaps the best example of all the option costs. If you are designing a dwelling you can optimise it to the existing energy costs but then let us see what option costs you might wish to incur. The case where the option costs are almost zero is loft insulation. You can optimise the thickness today and there really is no point putting in more insulation because when the price of energy goes up—if it goes up—you can just go up into the loft and put another inch in. On the other hand, the walls are at the other end of the flexibility scale. Once you have committed yourself to the U value of the wall, you have written off, to all intents and purposes, what you can do to that wall in the future, except at enormous refurbishing costs. In this sense the option costs rank economically the measures to which you ought to be paying most attention.

I think it is very unrealistic of us to ask the Department of Energy—'Supply us a cost profile to June 4th 1991 so that I can absolutely optimise my system'. To my mind that is bad design. Our design philosophy should reflect just how uncertain the energy situation is. The Department of Energy have supplied availability profiles up to I think, 1990, but they emphasise throughout the uncertainty in the range of estimates they are giving. If we really think that the buildings we are going to build are worth standing up for 60 to 100 years we really ought to be talking about fuel flexibility and an ability to handle the future situation, rather than, as it were, have spoon-fed prophecies handed down to us. An example which I think emphasises this point: if you took any economist back to 1965 and asked him whether he would, in fact, even with today's knowledge, have anticipated the present rise in oil prices he would still say that it would have been a low probability event because a whole number of factors would have had to coincide for it to be of that type of magnitude and to occur when it did. Yet, if you are putting up a building in some cases low probability events are still the ones you need to cover. Flexibility and option costs are things we have already got partially in our design philosophy when we put in stand-by generators but may be it is not a thing we think of when we think of putting up a building in terms of its wall and U values. Very simply if you have two walls of the same U value and one still has a vacant cavity, only the latter design leaves you the option in the future of filling it—It's as simple as that!

E. A. K. Patrick (British Gas Corporation, Research & Development Division): Before coming to my main point, may I refer to items that have arisen during this discussion? Firstly, may I say to Dr. Leach that we had indeed already picked up the need for lower-rated appliances for future housing. We are working on them, and they will be available.

Secondly, any fall-off in utilisation efficiency with intermittent use is largely a matter of heat capacity, and one might therefore reasonably expect that gas- and electric-wet central heating systems would behave similarly, and likewise, gas- and electric-air heating systems would behave similarly. In other words, to be fair one must compare like with like.

My original intention was to say a little about the information that Dr. Leach has given on page 174 where he gives comparative figures for a heat pump and a gas-heated wet central heating system. There are three points that should be made. Firstly there is the question of substitute natural gas. I was appalled to hear it suggested that natural gas will run out in 8–10 years', or indeed in the 30 years suggested earlier. Certainly a 100% substitute natural gas situation is unlikely to arise for, say, 40 years, and that is well into the life-time of any investment decision being contemplated today.

The second point is that the figures quoted appear to be for space heating only. The electric heat pump is credited with a coefficient of performance of 3 (an optimistic value) and debited with an electricity generation efficiency of 27%. The gas-heated system has been debited not only with the cost of making substitute natural gas, but in addition with a utilisation efficiency of 60%. If it were treated on a par with the heat pump, I would claim that it should be given a value of at least 75% because any shortfall from this figure must represent system losses, and the heat-pump system will equally be subject to these.

Lastly, there is some danger in considering figures for space heating only. Domestic hot water has also to be provided, and if, for example, the heat pump were to be associated with the rise of an electric immersion heater, the overall use of primary energy might well exceed that of a conventional combined gas-fired heating and hot water system.

Dr. S. Leach: I agree with everything Mr. Patrick has said.

B. P. Warwicker (Carrier Air Conditioning (UK) Ltd.): It has been said that Energy Conservation should only be implemented if it is cost effective. The people who control the purse strings will not, I suspect, be around when the energy deficit arrives. Therefore I submit that legislation or codes of practice are essential to override this pre-occupation with cost, to ensure a longer life and the continuity of fossil fuels.

I would like to take up a point made by Dr. Fisk about implementing something when it becomes cost effective. It may become cost effective, but there may not be any fuel or sufficient time left for the cost to be recovered.

Dr. J. Fisk: There is more than one meaning of the words cost and price. What Mr. Warwicker is actually suggesting, I beg to submit is that the price of energy in his opinion is still not reflecting its cost. I think Dr. Leach emphasised in the paper that in energy conservation we are always thinking about trading off one particular part of the economy against another. In fact

Perkins (see Pages 221–38) are the perfect example of this because as is mentioned, they are actually producing a product which saves energy. A year's production saves vastly more energy than the actual production consumes. In that sense we have a delicate balance where an energy saving industry is concerned. Remember that Perkins uses its own rate of return on capital which is a measure of how scarce it is to the Company. That capital would otherwise increase its own production facilities, and get more products to the market. I do not think anyone would suggest that in some sense we use that capital less efficiently to save a small amount of energy in the factory and in doing so lose a large amount in the production missed. You can push that argument right out to the whole economy. In a cost effectiveness analysis we are looking at how we are optimally to get this nation and the people in it, from our present situation to what we hope are their better and improved expectations in the future. It is rather important that we keep the same criteria of importance on all our economic frontiers. The suggestion that Perkins ought to cut back on its production in order to insulate, say, its roof will, if carried out, similarly in every other firm, effect the whole of the economy. That is a move that would really need to be justified.

P. Dey (*BICC Research & Engineering Limited*): I should like to make one point on the gloom and despondency that has been expressed in the conference on the availability of energy resources. It was said recently that the present petroleum and gas oils reserves are based on a 30% extraction rate and it was confidently expected that by the year 2000, technological advances would be made to make this extraction rate increase to perhaps 50% or more. That, I think, would extend the life of present North Sea oil reserves well into the middle of the next century or even to the end of the next century. So I do not think we ought to be too despondent about the future of the energy situation. Nevertheless, greater extraction rate would only give us a breathing space and we should not reduce the efforts on energy conservation measures we are discussing now.

9

Policies for Lighting Provision

G. P. CUNDALL

Synopsis

In working interiors lighting must enable the visual tasks which are to be undertaken to be performed quickly and accurately and it must contribute to a comfortable and pleasing visual environment. The main lighting criteria which have a significant bearing on energy consumption, and which need to be observed are discussed. These criteria can be fulfilled in most cases by providing virtually uniform illuminances using suitably chosen luminaires. Such solutions may not always be economical and it is suggested that task lighting and ambient lighting should be considered separately—particularly if daylight provides adequate ambient lighting. The importance of the contrast rendering factor in task lighting is emphasised and attention is drawn to a recent experiment in low energy office lighting. The importance of proper control and maintenance procedures is recorded and examples are quoted of the upgrading of lighting systems which have resulted in better lighting combined with reduced energy consumption. The paper concludes with a list against which proposed daylighting and electric lighting designs can be checked.

Introduction

In his autobiography, 'Slide Rule', Neville Shute quotes a definition of an engineer as 'someone who can do for ten shillings what any fool can do for a pound'. This definition is perhaps particularly relevant to the provision of light. Anyone—well almost anyone—can provide light by means of windows or electric lamps and luminaires but the scarcity and cost of energy now make proper design more important than ever before.

More people are motivated by the opportunity to save money than by exhortation so the object of any policy for lighting must be to obtain the greatest satisfaction for the least cost-in-use. In periods of crisis, such as during a strike which reduces electricity supplies, a policy of 'switching off something' may serve for the duration of the crisis—providing that switching off does not cause danger. This paper deals only with long term considerations. Although the scope of the paper is limited to working interiors much of it will be found relevant to other areas also.

Policies are needed for:

(1) the lighting of new buildings by day and during darkness.
(2) the control and maintenance of existing and new systems of day-lighting and electric lighting systems.
(3) evaluating existing systems and assessing the benefits of modifications or replacement.

There is not and never has been, a uniform policy for any of these subjects. The most coherent recommendations made in this country are those which the Illuminating Engineering Society has published in its Codes for Interior Lighting, in successive editions between 1936 and 1973. These Codes have had considerable and increasing influence in lighting design, but they do not control it. Perhaps it is because of this that misconceptions exist as to what the IES Code does recommend and what it does not. Policies are best made in the light of knowledge of the developments and experience which have led to the current recommendations for lighting design practice.

Developments in Lighting

In 1945 one of the commonest complaints in industry was of bad lighting. It was because of this that an intensive study of daylighting and electric lighting design was undertaken at the Building Research Station, as it was then known. Developments since then have been such that complaints about lighting are now few.

A large proportion of the buildings constructed before 1945 were of load bearing brickwork so that the sizes of windows were limited for structural reasons. If rooms had to be deep, as for example in many banking halls still in use, they also had to be high to admit a reasonable amount of light. Even so, daylight factors were often low—and were abysmal in rooms which faced 'light-wells'. In spite of this the use of electric lighting was unusual during the day because the amount of light it provided was so low. Such conditions were tolerated because expectations of lighting indoors were low.

Frame construction with wall cladding revolutionised daylighting opportunities. Wall to wall glazing could be provided—and often was. The increase in the amount of light it was capable of admitting by day was matched proportionately by the amount of light available from the tubular fluorescent lamp which was becoming increasingly popular. Although the efficiency of such lamps at that time was only about half of what they are now, they were 2–3 times more efficient than the filament lamps which had been the only suitable previous source of interior illumination. This improved efficiency was used to produce more light rather than to reduce energy consumption. That problems arose—with buildings which admitted so much light—is undoubted, but on the whole the greatly increased illuminance was enjoyed. Thermal and glare problems had to be solved, but not at the expense of returning to less than a desirable amount of light.

Lamp development was impressive. The efficacy of the tubular fluorescent lamp has been increased from 35 lumens/W when it was first introduced, to over 70 lumens/W now and its life has been extended from 2000 to 7500 hours. In the last few years metal halide lamps of efficacies of 85 lumens/W have appeared. Not only did efficacies improve, but lamp costs, relative to most other costs, fell so that electric lighting became more economical—right up to the end of 1973 when oil prices were quadrupled.

Quantity of light has never been the sole criterion by which to judge the goodness of lighting, but it is certainly one. Due to light becoming cheaper, successive editions of the IES Code have recommended the use of greater quantities. The increases recommended by the 1973 Code[1] however were modest and indeed, not all recommended illuminances were increased. The edition was perhaps most distinguished from its predecessors by the emphasis it gave to qualitative factors.

The Code has not been in use for long enough to have influenced a great deal of completed work. The lack of complaints about lighting experienced now therefore, must largely derive from the application of quantitative and qualitative criteria defined in previous editions.

Current Principal Criteria Affecting Energy Consumption

Criteria which have frequently been implemented in accordance with recommendations in editions of the IES Codes before 1973 which can have a marked effect on energy consumption are: illuminance, limiting glare index, colour rendering properties of light sources and colour appearance of light sources.

ILLUMINANCE

Until 1968, IES Code illuminance recommendations for task lighting were based on visual performance or the speed and accuracy with which things could be seen. The last Code published on this basis was that of 1961[2] which based its recommendations on the achievement of a visual performance of at least 90% of the maximum obtainable. In 1968 this basis was abandoned on the grounds that there was evidence that the benefits of good lighting were not limited by visual performance. Good lighting could help towards the creation of a sense of well-being from which the whole nervous system benefited.

Although experiments made in real work situations have validated increased productivity with improved lighting in industry and commerce, it has not proved possible to find a formula from which that illuminance could be calculated which would render the total cost of a particular process a minimum.

The 1973 Code therefore, states that its recommendations for illuminance should not be regarded either as minima or as optima, but as representative of good current practice. It will be realised that the higher the illuminance selected, the shorter will be the time that such an illuminance can be provided by daylight. For a given type of electric light source, the cost of energy for a unit of time will be virtually proportional to the illuminance selected.

GLARE CONTROL

Bright light sources in our normal field of view affect our ability to see or concentrate in several ways. Firstly, by the process known as phototropism the eye is attracted to the brightest objects in view. Secondly, the eye adapts to the average brightness (more strictly, luminosity) of the field of view and the usefulness of the lighting is therefore reduced if bright sources are visible. Their presence will always be a cause of some discomfort and eventually of fatigue, though they may not be consciously identified as such unless they are very bright.

A numerical scale of discomfort glare and the means of predicting it for uniform arrays of overhead luminaires was first introduced by the IES in its 1961 Code. Since then the design of luminaires for use in commercial applications in particular, has been changed notably. Few now permit any view of the lamps at normal angles of viewing and many manufacturers provide precise photometric data for their products so that likely glare can be assessed.

Light controllers and diffusers have often reduced the utilisation factors which could be obtained merely by the use of reflectors. For a given illuminance on a horizontal plane therefore, glare control can increase energy consumption. Sacrifice of glare control could reduce the value of lighting, however, by more than the value of the light absorbed in the light controller.

COLOUR

Colour appearance

The colour appearance of a source is defined in terms of its colour temperature in K. The colour appearance determines to a large extent the illuminance for which it is suitable. In his Presidential Address to the IES in 1973, A. H. Willoughby[3] summarised the relationship between illuminance and colour temperature in a diagram reproduced here in Fig. 1. This illustrates that the higher the colour temperature of the source, the higher the minimum illuminance for which it will be found to be acceptable. From this has been derived the following table of minimum illuminances for which lamps suitable for a wide variety of interior applications can be employed satisfactorily.

The table illustrates that it is possible to select a lamp of quite high luminous efficacy (subject to the fulfillment of other design criteria) down to illuminances of 75 lx which are below most of those required in working interiors. Lamps of the highest luminous efficacy—the metal halides—are only suitable for illuminances of 300 lx and over.

Colour rendering

The colour appearance of a lamp is not a guide to its colour rendering properties. The fluorescent lamps with the best colour rendering properties listed in Table 1 are Northlight, Colour Matching, Artificial Daylight, Kolorite, Trucolor and Natural. It will be noted that the minimum illuminance for which any of these lamps

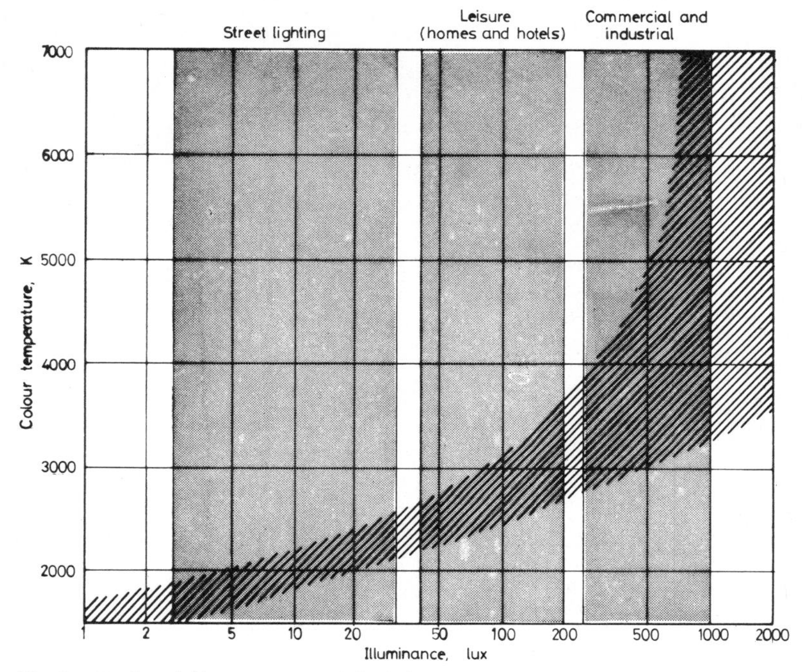

Fig. 1. Preferred illuminance ranges for light sources of different colour temperatures.

is suitable is 300 lx and for some, the minimum is 750 lx. The selection of a lamp with particularly desired colour rendering properties therefore, may entail a particular minimum illuminance.

Apart from the possible effect on minimum illuminance, however, it will also be noted from the table that the luminous efficacies of these lamps are significantly below those of the very popular general purpose White and Warm White lamps. The metal halide lamp makes a notable contribution to lighting design by combining good colour rendering properties with high efficacy.

For a given illuminance energy consumption will often be determined by the colour rendering properties required. For ambient lighting and some task lighting, however, there is evidence that, providing illuminances of 300 lx and more are maintained, equal satisfaction with the lighting will be obtained at a lower illuminance with good colour rendering lamps than with lamps of less good colour rendering properties.[4,5]

TABLE 1
LAMPS IN COMMON USE AND THE MINIMUM ILLUMINANCES FOR WHICH THEY ARE SUITABLE

Lamp type	*Designation*	*Range of luminous efficacy lumens/W*	*Colour temperature K*	*Preferred minimum illuminance Lx*
Tubular Fluorescent	Northlight Colour matching	35–50	6 500	750
Tubular Fluorescent	Artificial Daylight	25–35	6 500	750
Tubular Fluorescent	Daylight	55–80	4 300	400
Tubular Fluorescent	Kolor-rite Trucolor	35–55	4 000–4 200	300
Tubular Fluorescent	Natural	45–65	4 000–4 200	300
Hige pressure Discharge	Metal Halide	60–85	3 600–4 400	300
Tubular Fluorescent	White	60–80	3 500	200
Tubular Fluorescent	De Luxe Natural	30–45	3 600	200
Tubular Fluorescent	Warm White	60–80	3 000	75
Tubular Fluorescent	De Luxe Warm White	35–55	3 000	75
Incandescent	Tungsten	8–18	2 600–2 900	40
High pressure Discharge	High pressure Sodium	75–95	2 100	30

LAMP RATING

It will also be noted from Table 1 that for each lamp a range of luminous efficacies is given.

Generally speaking, the luminous efficacy rises with rating in lamps of a particular range. Design illuminance, room dimensions and other factors may determine the maximum lumen output of any particular lamp to be employed in a scheme so that it will not always be possible to employ lamps of the highest luminous efficacy in the range.

OTHER CRITERIA

Codes have recommended other lighting criteria and the 1973 Code gives specific numerical guidance on luminance distribution and modelling, but these do not necessarily affect energy consumption.

Energy Implications of Lighting

DAYLIGHT

Through any part of the fabric of a building, the greatest energy transfer, whether by means of conducted heat loss or radiated solar gain, takes place through the glazing. Heat loss through most kinds of single glazing is $5{\cdot}7\ \mathrm{Wm^{-2}\,°C}$. Heat gains depend on time of day and time of year, sky conditions, orientation, inclination, shading, transmission characteristics and the use of blinds. These variables are too numerous to summarise quantitatively in this paper. Proposed daylight design must take account of these energy transfers.

A. C. Hardy and P. E. O'Sullivan[6] have indicated that where the glazed area constitutes 75 % or more of the external wall, then the thermal environment will not be acceptable during periods of solar radiation, even with air conditioning. If glazing is restricted to about 45 % of external wall area, the room can be made acceptable with air conditioning. The amount of glass which can be used without incurring unduly high solar heat gains will depend on the weight of the construction. Thus the temperature in a building using a heavy weight construction system will rise less than in one using a lightweight construction system.

ELECTRIC LIGHT

It will be recognised that the energy implications of electric light are not confined to the energy which is consumed in its provision. At certain times the energy consumed by the lighting will add usefully to the heating requirements of a building and thereby help to offset other energy costs. At other times the lighting will produce an unwanted heat gain which has to be removed by expensive energy consuming cooling equipment. It follows that it is the use of electric light during daylight hours which has the greatest consequential effect on other energy costs as it is during these hours that cooling is most likely to be needed. Paradoxical though it may seem, it is during these hours that the highest illuminances from electric lighting are sometimes needed.

Current Design Practice

DAYLIGHTING IN FACTORIES

In single-storey factories it is theoretically possible to produce reasonably uniform illumination by means of roof lights, and they are commonly used. It is not always realised that the usefulness of this daylight may be reduced so much by the suspension in the roof space of such things as ventilation trunking, overhead conveyors and electrical bus-bars etc., that the electric lighting system has to be used all the time. When this happens all the penalties attributable to roof lights and continuously used electric lighting have to be paid together.

DAYLIGHTING BY MEANS OF SIDE WINDOWS

In side-lit buildings much current design practice acknowledges that electric lighting will be used to supplement daylight. It may be used to provide extra task lighting or to improve the balance of brightness across a room which has windows in only one side. The luminosity of the windows relative to other surfaces is then reduced—to the added comfort of the occupants—and the strongly directional properties of the daylight are beneficially modified. Millbank *et al.*[7] have shown that in offices the greater the illuminance available from electric lighting, the more it is used.

Acknowledgement of the problems caused by extensively glazed facades and the beneficial contribution which electric light can make during the daylight hours, is leading to the adoption of smaller windows. It is worthy of note that the long hallowed requirement of a 2% minimum daylight factor in school classrooms has now been relaxed by the Department of Education and Science. In order to

TABLE 2
DES GLAZING REQUIREMENTS FOR SCHOOL CLASSROOMS

	Maximum depth of room from outside wall			
	Less than 8 m	8–11 m	11–14 m	*Over* 14 m
Minimum proportion window area/ outside wall area	0·20	0·25	0·30	0·35

preserve a quantity of daylight and the ability to see out, however, minimum window areas are specified as shown in Table 2.

A good deal of work has been done to determine the occupants' preferred shape of windows of restricted area. Preferences depend to some extent on the nature of the view available, but in general it is found that a tall narrow window which gives a total cross-section of view from foreground to sky is preferred to the horizontally orientated shape. Tall, narrow windows are convenient structurally and also have the added advantage that the narrow shafts of sunlight they admit soon move so that occupants are not discomforted by prolonged subjection to radiation.

An alternative to the use of smaller windows is an external shading device. All too often such devices have had to be added after a building has been completed and they are not then always architecturally satisfactory. An example of a building where the shading is part of the architecture is the new GLC central office which employs external sun-blinds which are automatically controlled from photo-electric sensors on the roof.

LIGHTING DESIGN OBJECTIVES AND SOLUTIONS

The objectives of many lighting designs have been to

(i) provide adequate task illuminance for the purpose for which the space has been designed, over the whole working area
(ii) enable the space to be sub-divided in accordance with some agreed module, with minimum interference to services.

Windows have been designed and spaced to suit the building module and a regular array of overhead luminaires provided, often arranged to 'read' with the windows and spaced to respect the module in both directions.

Jay[8] has shown that by judicious choice of luminaires and decorations it is possible to design overhead lighting systems which will satisfy the principal criteria of task illuminance, luminance distribution, glare control—and modelling too when there is a reasonable amount of daylight to provide a horizontal component of illumination. There can be no question but that systems designed on these lines have given considerable satisfaction and have much to commend them in convenience and simplicity.

CRITICISMS OF THE SOLUTION

Although the smaller window has made a significant contribution to energy saving, the electric lighting is sometimes considered to be extravagant in its own consumption and in the cooling plant it calls for. This criticism requires examination on the basis of the established criteria, the principal one in relation to energy, being illuminance.

TASK ILLUMINANCE

The IES Code lists over 470 numerical illuminance recommendations of which 95 exceed 500 lx. Of these 95, 60 are specifically advised to be provided by local or localised lighting. Of the remaining 35, localised lighting may be judged to be suitable in many cases. Recommendations for overall illuminances exceeding 500 lx are therefore few; indeed not all the recommendations for illuminances up to 500 lx need necessarily be provided by uniform lighting. Of those recommendations which do exceed 500 lx, perhaps the most controversial will be; deep-plan offices, business machine and typing offices, and drawing offices, all of which are recommended to be provided with 750 lx. Apart from a few industrial areas in which very exacting work is done, no other space is recommended to be provided with a general illuminance of more than 750 lx.

Ambient illuminance outside daylight hours

Satisfactory ambient illuminances are related to task illuminances in a way to be described later, but it should generally be at least 150 lx.

Ambient illuminance during daylight hours

In a room which admitted but a small amount of daylight, or none at all, 150 lx would seem exceedingly inadequate by day, as one's expectations of lighting are high.

If there are windows which do not light the whole room adequately, then the electric light used to supplement the daylight must be sufficient to create a satisfactory luminance balance. Hopkinson and Longmore[9] have explained the procedure for calculating the satisfactory supplementary illuminance in side-lit rooms. The figure depends on the brightness of the sky seen from the supplemented area and the daylight factor, but will often be found to be around 400 lx or more. Where there are no windows at all, the IES Code recommends that an illuminance of 500 lx is provided to compensate for lack of daylight.

On the basis of ambient requirements in spaces which receive restricted amounts of daylight, the provision of task illuminances up to 500 lx by means of a uniform lighting array providing an ambient illuminance of the same order, would appear to be justified, in many cases by present concepts. In drawing offices, typing and business machine offices it is possible to use localised lighting—that is lighting which provides a high illuminance over working areas and at the same time provides adequate but somewhat lower illuminance elsewhere. The provision of 750 lx in deep open plan offices then becomes the principal exception to the general maximum of 500 lx which appears to be a reasonable figure on the basis of accepted criteria.

There have been many office and shop installations in which the illuminance provided exceeds 500 lx, or even 750 lx, but it should be noted that such higher illuminances are not currently recommended by the IES. If savings are to be made without reducing the standards which are recommended, the presently used criteria will require examination and perhaps different methods of design will need to be adopted.

Other Approaches to Lighting Design

A working interior must be provided with:

(a) adequate lighting for performing the visual tasks to be undertaken, quickly, accurately and without fatigue.
(b) comfortable ambient lighting which will enhance the appearance of the space.

Reference has already been made to the use of localised and local lighting. Localised lighting is designed to illuminate an interior and at the same time provide higher illuminance over a particular part or parts of the area. Local lighting is designed to illuminate a particular small area which does not extend beyond the area of the visual task.

Hopkinson advocated many years ago that task lighting and building or ambient lighting should be considered separately. One reason for their so often being combined into a single system has been explained in the preceding section. Another reason has been the fear that their separation at 500 lx or less would fail to meet the second criterion of enhancing the appearance of the space. The need to conserve energy merits reconsideration of the provision of separate task lighting and building lighting systems.

TASK LIGHTING

The manner of providing task lighting itself requires examination. It is a matter of common experience to have been unable to read a printed page or words on a chalkboard because of the reflection of a window or luminaire on the surface. In these circumstances, the usefulness of the light incident on the surface is entirely counteracted by a veiling reflection. The visibility of a task therefore, depends not only on the measured illuminance but on the ability of that illuminance to reveal differences of contrast between detail and background. This contrast ratio is defined in CIE Publication No. 19[10] as Contrast Rendering Factor.

The Contrast Rendering Factor (CRF) is unity when the task is placed in an integrating sphere where the light on it comes from all directions. A large uniformly diffusing ceiling will give a CRF of approximately 1·0. If that part of the ceiling which would be reflected in a mirror in the plane of the task were blacked out, the CRF would be greater than 1·0 as there would be no veiling reflection at all. Since no surface is absolutely matt there will always be some reduction in visibility due to veiling reflection unless light sources are so placed that such reflections are avoided.

Clearly, under normal lighting conditions, CRF will vary from place to place and according to the direction in which the task is viewed. Values of CRF have been determined for pencil handwriting viewed at different angles with a variety of lighting systems. For typical overhead office lighting systems they fall in the range 0·8 to 0·9. If luminaires with polarisers or with optical systems giving 'batwing' distributions are used, the CRF's fall in the range 0·9 to 1·1.

Since the visibility of a task is not a function of its illuminance only, the American IES has started to specify certain lighting requirements in terms of Equivalent Sphere Illuminance (ESI), *viz:* to require that the lighting shall provide the same visibility as that obtained with the task in an integrating sphere with the specified illuminance. If the CRF is unity then the task illuminance and the ESI will be the same. The relationship between CRF and task visibility is such that if the CRF should be only 0·9 the task illuminance will have to be twice the ESI to provide the same visibility. Referring to the previous section, therefore, if a conventional office lighting layout provides 500 lx with a CRF of 0·9, the same visibility could be obtained from a lighting installation giving only 250 lx and a CRF of 1·0.

The ratio of the equivalent sphere illuminance to the illuminance provided is termed the Lighting Effectiveness Factor (LEF). Lighting giving a CRF of 0·9 will produce an LEF = 0·5. If the lighting can be so arranged to give a high LEF then significant energy saving should be possible in task lighting.

The application of a low energy approach to office lighting. Using these concepts, Cuttle and Slater have reported on a study of office lighting[11] with aims similar to those stated at the beginning of this section. Part of the study is summarised here.

The office was occupied by about fifteen people. Each office worker was provided with a 600 mm 20 W fluorescent lamp mounted in a metal reflector, 500 mm above each end of his desk. The angle of the reflector was adjustable within limits designed to ensure that distraction was not caused to other people. With this arrangement veiling reflections of these sources were absent from the control area of the desk, at which position the illuminance achieved was 280 lx planar. The CRF, however, was 1·11 and the LEF 2·04 so that the ESI was 570 lx.

In addition to task lighting a low level of general lighting was provided by night, giving an average of 130 lx. This raised the desk top illuminance to 410 lx planar and to 740 lx ESI.

The room in which this experiment has been performed has windows on three sides and an unusually high average daylight factor of 3 %. The office was previously lit by overhead fluorescent luminaires giving 600 lx. They were usually used all day in spite of the good daylight factor.

Now, each worker decides for himself whether or not he requires his desk illuminance to be increased and he switches on and off as he pleases. The low level background lighting is controlled by a photocell. Statistical analysis of ratings of the appearance of the room with the original conventional lighting and with this lighting, do not indicate any dissatisfaction. Individual members of the office are often out and their desk light is switched off when they are out. The saving in energy consumed by lighting over a half year period has been 82·8 %.

This remarkable saving was largely achieved because in the original scheme, the ambient lighting of the whole space was raised to provide adequate task lighting, even though daylight provided adequate ambient lighting. The very good daylighting in the room makes it too untypical to be sure about the variety of interiors in which such a scheme would be likely to be found acceptable. The authors would acknowledge that the appearance of the experimental desk fittings requires to be improved by expert design.

Apart from the extremely high direct energy savings, the system has the following advantages:

(a) the summer cooling load would be reduced enormously

(b) the worker controls his own luminous environment
(c) when he uses his desk lighting, his work is lit preferentially in respect of everything else so that concentration is encouraged. His task visibility is better than it was.

AMBIENT LIGHTING

The kind of electric lighting to be provided for ambient lighting may be governed by the quantity and quality of the daylight available in the space. Where the daylight is good it will provide satisfactory ambient lighting much of the time. The electric ambient lighting will therefore only be used during hours of darkness. It will not be necessary for it to provide a uniformly high illuminance. Where the daylight has to be supplemented, as in deep spaces, or where it is excluded altogether, explanations have been given for the practice of providing general illuminances up to 500 lx. It is necessary to consider whether there is an opportunity for energy conservation in both of these situations but especially in deep or windowless spaces.

Luminance distribution

It has been established that to ensure comfort the ambient luminance should not be less than one-tenth of the task luminance, and then only if there is an immediate task surround luminance of intermediate value. Otherwise, the environmental luminance should be about a third of the task luminance.

Since luminance is a product of illuminance and reflectance, surface colours are at once seen to be important in establishing acceptable luminance ratios. Light colours on large surfaces will clearly help to provide the required luminance with minimum illuminance, but it is necessary to avoid the creation of too bland an environment. A successful interior will not just respect the desirable luminance ratios, it will provide some interest and stimulus.

If only a modest general lighting system is employed it will often be possible to raise the average luminance by selective lighting. Thus walls which may be interestingly decorated or textured, can be preferentially lit. Other objects of interest can be highlighted by means of spot or flood lamps. The scope available will obviously depend on the size and nature of the space to be lit. Whilst it is important to avoid gimmicky or over-contrived solutions which will pall with time, it will be found that more interesting as well as more economical ways can be found of creating a good ambience than by merely flushing the space with light. Indeed such a practice can fail hopelessly. It failed in a factory known to the author where, although the ambient illuminance is 500 lx, the effect is gloomy because the ceiling space above the luminaires is painted black and all walls are in the darkest possible brickwork in which even the mortar is black.

The application of these principles to spaces requiring permanent artificial lighting—supplementary or otherwise—is worthy of study. Ambient luminance is obviously determined to quite a large extent by the luminance of vertical surfaces. It appears at least possible that ambient illuminances in such spaces should be specified in scalar rather than planar terms. The calculation currently used for determining the optimum supplementary illuminance gives a result which is directly proportional to the luminance of the sky as seen through the window from the area to be supplemented. Guidance would be useful on the extent to which the calculation will give reliable results if the glass is tinted. By reducing the daylight factor, tinted glass will increase the area requiring supplementary illumination but might it reduce its level? It would also be useful to know the maximum absorption in the visible spectrum which is acceptable for window glass.

SOME IMPLICATIONS OF SEPARATING TASK LIGHTING FROM AMBIENT LIGHTING

It has already been explained that task lighting may be provided by either localised or local lighting. The sources should be so placed as to avoid veiling reflections. Local lighting will usually be required to be mounted on the machine or furniture. Industrially, an electricity supply is provided to every machine but in an office, each desk is normally not so provided. If long flexible trailing cables are to be avoided extensive provision must be made in floors and skirtings for socket outlets. Such trunking systems are expensive to install but their cost need not all be allocated to the lighting system. More and more office workers use telephones or other even more advanced communication aids, mini-computers, or simpler calculating machines, electric typewriters, erasers, etc., and it may well be that the growth of such aids alone will merit the provision of such distribution facilities.

Control and Maintenance of Lighting Systems

CONTROL

Over the last fifteen years or so the practice of switching lighting in large blocks has become more common. In large spaces it is certainly difficult or inconvenient to arrange for numerous switching positions. Control of overhead general lighting requires consideration in three respects concerning energy conservation. Firstly, areas which receive adequate daylight for much of the time should be switched separately from those which are needed all the time. Secondly, the sizes of lighting blocks should be related to work areas. Thirdly, if high general illuminances are provided, provision should be made for reduced lighting to be available for cleaners, etc. Photo-electric cells should be considered for controlling the ambient lighting over normally daylit areas.

MAINTENANCE

Dirt is the enemy of both daylight and electric light. It obviously follows that windows, lamps and luminaires should all be kept clean. The problem is to ascertain what is a reasonable cleaning interval. IES Technical Report No. 9[12] shows that the cost of light is minimised if cleaning takes place when the cost of light absorbed by dirt becomes equal to the cost of cleaning. Thus optimum cleaning intervals can be calculated or tables in that report can be used as a guide to reasonable intervals for different circumstances.

Evaluation of Existing Systems

All existing electric lighting systems should be checked to see whether a profitable investment could be made by their replacement. It is not only filament lamp luminaires which might merit replacement; an example to be given later in this section, illustrates that even tubular fluorescent luminaires may merit replacement in some circumstances.

The Lighting Industry Federation is to be complimented for emphasising the importance of using the most efficient lighting equipment by introducing an 'Energy Management in Lighting Awards Scheme'. Entries for this scheme cannot be quoted at this stage, but the Lighting Industry Federation has provided examples of the results of changing some existing lighting schemes. The information provided is insufficient to enable full cost-in-use studies to be made but the savings claimed appear to be notable enough for the absence of additional information to be of minor importance. For the sake of brevity, the information given is tabulated.

TABLE 3
EXAMPLES GIVEN BY LIF OF THE RESULTS OF CHANGING EXISTING LIGHTING INSTALLATIONS

Application	*Lamp types*		*Change in illuminance*	*Savings*		*Estimated 'pay-back' time*
	Initial	*New*		*Load*	*p.a.*	
Steelworks	1000 W Tungsten	600 W HP Sodium	+358%	22·4 kW	£855	18 months
Foundry	400 W + 500 W Mercury and Tungsten blended	360 W HP Sodium	+ 31%	46·5 kW	£2843	Not stated
Alloy works	400 W Mercury	310 W HP Sodium	+ 50%	−20%	£6000	Not stated
Textile and Lace–Security Lighting	500 W Tungsten	250 W Mercury MBF/U	None	5·5 kW	£398	3 months
Hotel corridor	100 W Tungsten	20 W Fluorescent	None	2·6 kW	£400	5 months
Power Station	Fluorescent 840 off	HP Sodium 92 off	None	86 kW	Not stated	Not stated

Whilst still following the traditional design concepts, these schemes illustrate the considerable benefits to be obtained by the use of efficient and up to date equipment. There must be numerous corridors and toilet areas in offices as well as hotels and other buildings which employ filament lighting throughout the working day in which considerable savings could be made by providing better fluorescent lighting.

Conclusion

It is hoped that this paper will have illustrated that there are more ways of saving energy on lighting than by merely 'switching off something'. The fact that the energy consumed by lighting appears to be only around 13 % of all electricity consumption and 3 % of total energy consumption is no reason why lighting should not be made as energy-effective as possible—particularly since it may be responsible for consequential energy consumption by air conditioning plant.

There are so many ways of lighting buildings, particularly by day, that any summary of recommendations runs the risk of being inadequate or an over-simplification. For working interiors, however, the following may be a suitable check list.

(1) In single storey buildings in which roof lights are being considered check:
   (a) the effects of heat loss and solar heat gain through the glazing on the total energy requirements of the building;
   (b) the extent and location of overhead services and equipment which may be used which could affect daylight distribution and its consequent usefulness.

(2) In side-lit buildings:
   (a) unless the view to be obtained from within is outstanding, be sparing in the use of glazing, particularly on walls which do not face north;
   (b) if the view to be obtained is of such value that large windows should be employed, install shading devices such as automatically operated blinds;
   (c) make the most of the glazing used; consider spacing windows and providing splayed reveals so that
      (i) the windows appear as large as possible when viewed from inside,
      (ii) the maximum of light flows into the building with the minimum of dark spaces between windows
   (d) consider the use of tinted glass.

(3) In electric lighting design:
   (a) consider separately task and ambient lighting systems—particularly where the provision of only one lighting system will impose high summer daytime loads;
   (b) provide task lighting of the standard specified in the IES Code but do not provide more;
   (c) never provide light for the sake of providing heat;
   (d) endeavour to design task lighting to provide a high Contrast Rendering Factor;
   (e) ensure that the ambient lighting in conjunction with the

decoration provides an acceptable luminance distribution. Consider its automatic control;

(f) consider the use of some emphasis lighting in combination with a lower general illuminance. In so doing maintain a natural simplicity of design without 'gimmicks'.

(4) In the wiring of electric lighting systems ensure that lights can be controlled:

(a) in accordance with work areas;

(b) in accordance with the amount of daylight available. Consider the use of photo-cell control. This will not only save lighting energy but air conditioning energy too—or lead to greater thermal comfort.

(5) Ensure that windows and all other equipment are kept clean and well maintained. Ensure that someone in the organisation is responsible for this.

(6) Keep up-to-date with lamps and lighting equipment. Check the cost-in-use of existing equipment against modern equipment.

Acknowledgements

The author acknowledges with thanks:

The Illuminating Engineering Society for permission to reproduce Fig. 1. The Lighting Industry Federation for permission to quote the examples listed in Table 3. The Pilkington Advisory Bureau for advice and information provided on the installation described in the section entitled 'Task Lighting'.

REFERENCES

1. 'IES Code for Interior Lighting'. Illum. Eng. Soc. (London) (1973).
2. 'IES Code for Interior Lighting'. Illum. Eng. Soc. (London) (1961).
3. Willoughby, A. H. (1974). New lamps: their suitability for interior lighting. *Lighting Research and Technology*. **6.1.**
4. Bellchambers, H. E. (1971). *Illumination, colour rendering and visual clarity*. CIE, Paris. p.71.25.
5. Rowlands, E., Loe, D. L., Waters, I. M. and Hopkinson, R. G. (1971). '*Visual performance in illuminance of different spectral quality*'. CIE, Paris. p.71.36.
6. Hardy, A. C. and O'Sullivan, P. C. (1967). *Insolation and Fenestration*. Oriel Press Newcastle upon Tyne.
7. Millbank, N. O., Dowall, J. P. and Slater, A. (1971). Investigation of maintenance and energy costs for services in office buildings. *J. IHVE* 39.145.

8. Jay, P. (1968). Inter-relationship of design criteria for lighting installations. *Trans. Illum. Eng. Soc.* 33.47.
9. Hopkinson, R. G. and Longmore, J. The permanent supplementary artificial lighting of interiors. *Trans. Illum. Eng. Soc.* **24** (3) 121–139.
10. *A unified framework of methods for evaluation of visual performance aspects of lighting.* (1972). CIE, Publication No. 19.
11. Cuttle, C. and Slater, A. I. (1975). A low energy approach to office lighting *Light and Lighting*, 68.20.
12. 'Depreciation and maintenance of interior lighting. Technical Report No. 9. Illum. Eng. Soc. London 1967.

Discussion

J. B. Collins (*Building Research Establishment*): I should like to support Mr. Cundall in his views on the importance of treating task lighting and building lighting separately. He has shown us a number of examples of how the wattage can be reduced in this way, and it is my personal view that one might set targets in terms of wattage per square metre for lighting and leave the lighting engineers to compete to provide us with the best lighting schemes in terms of visibility of the task and of the satisfaction with the visual environment within these wattage targets. I should like his comments on this point.

Another point I should like to make is in connection with Mr. Cundall's point of control systems and possible photoelectric control of lighting. We do now have some data of the way in which lighting is used throughout the day at various times of the year in various buildings—one example is illustrated in Fig. A. The two graphs indicate the amount of lighting actually switched on during the day, both in winter and in summer. You can see there is actually very little difference in the use of lighting during the daytime between winter or summer, which does seem to indicate that more lighting is being used than is necessary during the summer, and in fact if you can adjust the amount of lighting used by sufficiently sophisticated photoelectric controls, you can reduce it to no more than the necessary value. The size of the shaded areas in relation to the total give you some indication of the amount of energy you might save. We are looking into the cost effectiveness of these control methods because they are fairly expensive at the moment. Some of the use shown occurs at night at times when the area is unoccupied, and there is obviously quite a possibility here of saving by use of appropriate time switches.

I know of an installation in Denmark which shows that dimming controls are in fact in use and practicable. The installation has been working quite satisfactorily providing just the required amount of artificial light to top up the daylighting which comes through roof lights. The artificial lights are adjacent to the roof lights and are controlled by photoelectric dimming which provides artificial lighting just sufficient to keep the illuminance up to about 700 lx.

Various speakers have mentioned the importance of window size. We have done some experiments at the Building Research Establishment to obtain an

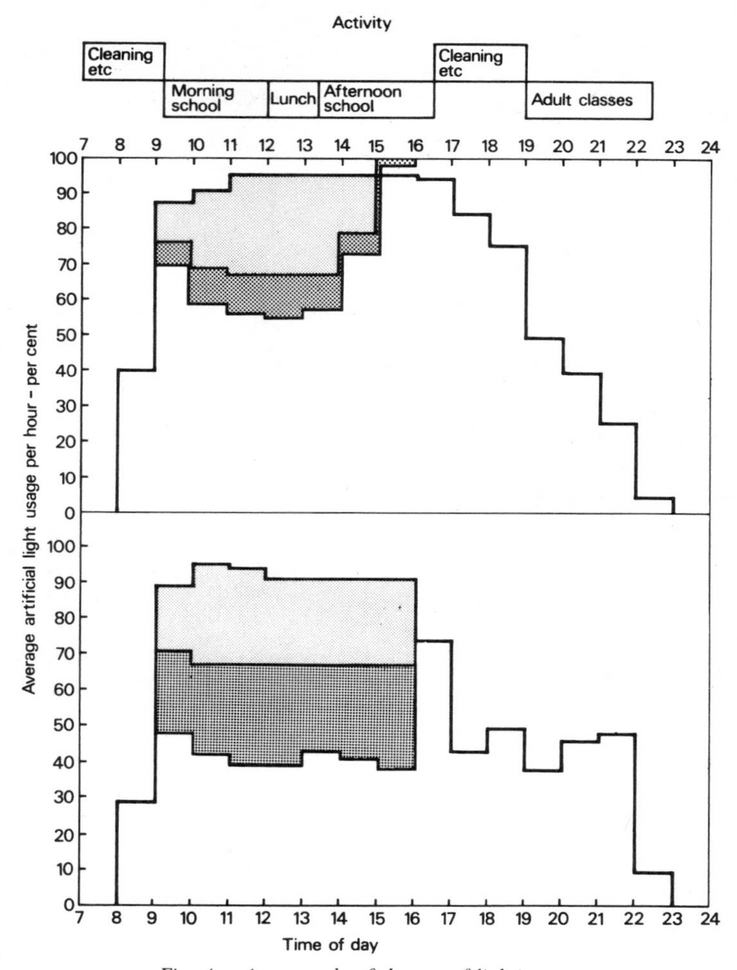

Fig. A. An example of the use of lighting.

idea of preferred window sizes and we do find that a window of about 25 %–30 % of the external wall area is in fact the most popular. Fig. B is a view of the model used for the study, with the type of window which was judged to be the most satisfactory by the people who had the opportunity to adjust not only the size of the window but also its shape. The boundaries could be adjusted independently. People do not like to see too much sky—they do like a more extended view of the outside and in contrast to the windows advocated by Mr. Cundall, our subjects tended to prefer windows that are rather lower and have an average ratio of about 0·4 for height to width. What indeed people do need to see through their window is a view of the skyline and it is thought that this is important. I do realise that the tall narrow windows are more advantageous structurally, and they probably have an advantage in terms of the movement of sunlight about the room—that it is not on one person too long, but there are other considerations in the design of windows. However, we need not have too many inhibitions about reducing the size to around 30 % of the facade.

Fig. B. A model for assessing preferred size and shape of windows.

K. R. Ackerman (British Broadcasting Corporation. President—Illuminating Engineering Society): Mr. Collins is Chairman of the Illuminating Engineering Society panel working on a new IES. Lighting Code which will be published early in 1977. This new edition will be taking particular account of the energy situation. Mr. Collins also played a leading role in the production of the current 1973 IES. Lighting Code.

J. Vollborth (Haden Young Ltd.): How seriously does Mr. Cundall expect us to take the low energy approach to office lighting using the localised luminaire? This seems to be going back into the dark ages literally. The system has become more intricate by giving it expressions like contrast rendering factors and lighting effectiveness factors but how do we convince a customer that if he is reading 250 lx on his meter it is really equivalent to 500 lx? As far as designers are concerned there is very little information published to help them achieve these contrast rendering factors which seem to be empirical anyway.

In the paper, lighting was quoted as 13 % of electrical energy used and 3 % of total energy used, I would question these figures as being low. Recently I arrived at the figures of 15% for lighting as a proportion of total electrical energy and 4·2% of total energy. There may not seem a lot of difference in these figures but 2% difference is in fact equivalent to about double total public lighting consumption or £60 million in cost.

G. P. Cundall: The experiment quoted in 'Task Lighting' was made in an office in which the average daylight factor was 3%. The ambient illuminance was therefore satisfactory for much of the working day. The work in the office could be done satisfactorily if only the task illuminance was supplemented. This is why the savings were so striking. Of course the experiment must be taken seriously: When such savings in electrical energy for lighting alone can be made without causing dissatisfaction, such an experiment cannot be dismissed. If such a solution could be applied to a building which required cooling, significant savings in cooling energy would also be obtained.

To say however that the experiment is to be taken seriously is not to say that the method used will be universally applicable. In deep buildings with restricted daylight a supplementary ambient illuminance of 150 lx may not be satisfactory. But as many authors have reminded us, most buildings are not new. The cost of energy and our moral responsibility to husband it, requires us to review the standard solutions to lighting which have been adopted over the past 15 years or so and this experiment serves at the least as a challenge to lighting designers.

The figures given for the consumption of energy for lighting are taken from an Electricity Council publication 'Maintain good lighting and still save energy' Ref. EC 3168. Even this source of information must rely on estimates because separate metering of energy for lighting and other purposes is not common. If other estimates indicated percentages of 15 where I have quoted 13, or 4·2 where I have quoted 3, I would submit that the magnitude indicated remains of the same order.

C. Cuttle (Pilkington Bros. Ltd.): I would like to refer to Mr Vollborth's question of 'how seriously can we take the type of low energy lighting illustrated by Mr. Cundall?' To evaluate lighting on the basis of the number of lx it puts onto a task is no more sensible than evaluating heating solely by the kW output of plant. As the appropriate criterion for heating is comfort, so the appropriate criterion for task lighting is task visibility, this being most conveniently specified in terms of Equivalent Sphere Illuminance. As for all lighting measurements, ESI is related to the human visual response and so is not a purely physical unit, but it is misleading to describe it as being empirical. It is a measurable quantity that defines the relative visibility of a task.

Finally, if the slides left Mr. Vollborth with the impression that this lighting approach represents a return to 'the dark ages', I am pleased to extend an invitation to him to visit the installation in St. Helens, as I feel sure that this would convince him that this is not so.

10

Insulation of the Housing Stock—a National Problem

TED NICKLIN

Synopsis

Our stock of 18·5 million dwellings is old and largely uninsulated. It is grossly inefficient in its use of energy. A substantial proportion of the housing stock falls into a limited number of types in respect of construction, size, and plan form and presents very few technical problems in improving its thermal performance sufficiently to provide comfort conditions with the consumption of substantially less energy. The problem becomes confused by the very wide range of standards of heating throughout the housing stock and by the variety and diversity of the population. All the same the installation of 7 million central heating systems between 1960 and 1972, with an increase of 18% in the number of dwellings in the same period with no increase in nett energy consumption demonstrates that substantial improvements and economies can be achieved in this field. The proposal is for an inversion of domestic energy tariffs to provide a small quantity of cheap energy and any further consumption at a high cost, with very substantial grants for thermal upgrading, both insulation and improved installation, including the removal of all electric heating. The capital cost would be more than covered by the cessation of further investment in electrical generation.

Introduction

The median age of our housing stock is almost fifty years, and, given the current rate of replacement, most of our houses will be well over a hundred years old before they are demolished. There is little reason to suppose that the rate of house construction will increase significantly for some time to come, and we can therefore conclude that our existing housing stock is here to stay well into the next century. The energy it consumes for space heating alone, quite apart from water heating, light and power, accounts for about 20 % of our total energy consumption. This is more than we consume for all forms of transport, and is second only to industry in the rank order of energy consumption in this country.

We have a long tradition of not taking heating seriously, especially in our houses. Our climate has seldom been rigorous enough to freeze us to death and some of our forebears seem to have found a strange kind of virtue in not being too warm. For others there was abundant cheap coal, followed by abundant cheap oil. As a result only about four or five million of our 18·5 million dwellings have any insulation at all; for the greater part of our housing stock you could only reduce the thermal performance of the building fabric by removing some walls or roofs altogether.

Now, two powerful new factors have emerged. Increasing numbers of us now expect to be as warm at home as we are at work or in any public building, and at the same time energy costs are on a dramatic upturn. It seems obvious that here is an open and shut case for a massive conversion of our housing stock to a high level of thermal efficiency. And yet, two years after the October war, nothing has happened. Why is this? Let us look more closely; first at the houses themselves, for most of these are quite easily classified and have, by and large, a remarkable uniformity of construction and even shape. Secondly we must look at the people in the houses, and it is here that we find the complications begin to emerge. For lo! the problem is a social one, not simply a technical one, which means that it enters a wholly new order of difficulty. One suspects that even if there were the political will to grasp this seemingly wide open opportunity for energy conservation, the administrative resolve would falter, faced with the human messiness of it all.

Profile of the Housing Stock

About 4·5 million of our 18·5 million households are in flats and detached houses which are not readily classified, but the remaining 14 million 'ordinary houses' can be divided into four distinct categories according to age:

Pre 1919: about 5·5 million, almost all with solid brick external walls, pitched roofs, suspended timber floors, areas varying over a range from 60 m^2 to 150 m^2, high ceilings, small windows, no insulation. The characteristic plan is the narrow fronted terrace with a rear offshoot.

1919 to 1945: about 3·5 million dwellings, the most uniform group, with cavity walls, pitched roofs, suspended timber floors, areas varying only a little around 90 m^2, 2·4 m ceilings, still modest window areas, no insulation. The spec builder's standard plan semi predominates.

1945 to 1965: about 3·0 million houses, still largely traditional construction with cavity walls and pitched roofs, but an increasing use of solid ground floors, substantial increases in window areas, with floor areas very uniform at around 80 m^2 and, with the adoption of 2·2 m ceilings, minimum volume. Still generally no insulation.

1965 to present day: about 2 million houses, with more varying forms of construction including some flat roofs, and external walls which although predominantly cavity brick/block construction, include substantial areas of various forms of panel construction. Areas and volumes very uniformly minimum but windows getting even bigger. The most significant change at 1965 is that, for the first time, thermal insulation is required by law. The standard demanded is little more than a nominal improvement on the construction used for decades before, but it is nevertheless a significant change. This is a rough and ready classification, and I am certain every member of this audience lives in an exception, but I believe it illustrates the general shape of the problem, so far as building form and fabric is concerned.

Improving the Fabric

Technically there is no mystery about how to improve the insulation and reduce air infiltration in most of our housing stock, and consequently in the houses that are adequately heated and equipped with reasonably efficient heating systems, there is plenty of scope for significant reductions in energy consumption.

There is no need to go over them in detail, but a number of points do need to be made. First roofs; if God had not made pitched roofs it would have been necessary for the mineral fibre insulation industry to invent them—no problem. Flat roofs are the rare exception in housing and this is fortunate because they present one of the most difficult problems. There is rarely sufficient spare live load capacity to allow the use of what I believe to be the ideal solution—the upside down roof, consisting of slabs of closed cell polystyrene laid dry on top of the waterproof membrane, held down by pebbles or precast concrete slabs. Otherwise the only alternatives are a layer of insulation on top, with a waterproof covering—an expensive and tricky operation—or internal soffit lining of foamed polyurethane with integral flame retarding surface. This last may be one of the few situations where warnings of interstitial condensation may in my opinion have some substance. I would not use an insulating lining in a bathroom or kitchen without taking special precautions to prevent possible damage to the fabric, but elsewhere, I believe the 'condensation bogey' is based on assumptions of steady state atmospheric conditions that rarely, if ever, arise in housing in this country.

Cavity walls can be filled with urea formaldehyde foam or mineral fibre, and now that the Department of Environment's left hand has found out what its right hand is doing, we can do it legally. There is one proviso I would make about this method; in the case of intermittently heated spaces it may be less effective than an internal lining, because, due to the thermal capacity of the inner block or brick skin this will take time and energy to warm up to a comfortable temperature.

In any case, in pre 1919 houses, and there are 5·5 million of these, there is no alternative, if wall losses are to be reduced, to a lining such as expanded polyurethane battened or stuck to the inside face of the walls. This is likely to be a disruptive operation, but apart from gables in ends of terraces and semi-detached houses, of which there are not a high proportion in this age group, the external wall area in these generally narrow fronted terraces is fairly limited.

Double glazing of existing windows cannot be justified in economic terms, but heavy curtains or, better still, insulating shutters make a great deal of sense. They can be drawn for sixteen hours out of twenty four in the coldest weather, and in the case of windows facing the sun, solar gain can more than offset the daytime losses, even in winter. A practical, light, cheap, insulating shutter, as a marketable piece of hardware at about a quarter the cost of double glazing, is waiting for an inventor and an entrepreneur.

Provided the ventilation under suspended timber floors is no more that the trickle necessary for the health of the timber, it is probably not worthwhile adding insulation. The U value will be about $0{\cdot}65\ \mathrm{W/m^2}$ for a terrace house and $0{\cdot}85\ \mathrm{W/m^2}$ for an end of terrace or semi-detached. Corresponding figures for a standard solid concrete ground floor with damp-proof membrane are 0·45 W/m^2 and 0·65 W/m^2. Most houses are draughty and, with involuntary air change rates of 2 to 3 per hour, many lose as much as 30 % of their heat through air infiltration. While draught stripping and sealing up cracks are cheap and easy, the domestic rigours of controlling child/dog induced air change may be an altogether different order of problem, but lobbies, door closers and rising butt hinges can help considerably even here.

A device that has produced 10 % savings on heating energy is a small fan installed in the ceiling above the top landing, drawing air from the roof space. Very frequently even in winter the air in the roof space is warmer than the air outside as a result of solar radiation, and provided the fan is thermostatically controlled, it provides a marginal internal rise in air temperature as well as reducing infiltration of cold outside air by raising internal air pressure.

Heating Standards

The effectiveness of such measures in reducing energy consumption depends on the standard of heating—no heat, no saving—the more heat, the more saving. But now assume a very modest level of heating in, for example, an inter-war semi-detached house of 96 m^2 volume 240 m^3, and taking one third of the volume to be heated for 16 hours a day to 18 °C and two thirds to 10 °C; and the whole house heated to 10 °C for 8 hours a day—say 1000 degree/days, which is about half the total that would follow from heating to Parker Morris standards. In such a house a reduction of over 40 % in energy consumption can be achieved by three measures alone:

Reduce air change from 2 to 0·75 per hour by simple draught stripping.
Install 50 mm glass wool mat in roof.
Inject urea formaldehyde foam into wall cavities.

The house I am citing would have an annual energy loss, before insulation, of 46·5 GJ which is equivalent to a demand of 66 GJ assuming an installation of 70 % efficiency. This is just about the national average nett demand of 700 therms.[1] The capital cost of the three conservation steps all done by a contractor, without any 'do-it-yourself' element, will be about £150. Analysis of typical houses from all the categories outlined produces figures very much in line with the example quoted;[2] provided the house is heated to even a modest level, reductions of the order of 40 % in annual consumption are easy to achieve, at a capital cost that will show repayment in a few years at present tariffs.

Peak load, when external temperature drops to 0 °C will be reduced from 6·8 kW to 3·4 kW, a 50 % reduction. Thus if the house is electrically heated, the capital cost of halving this peak demand on the electricity supply system will be less than £50 per kW. Consider the significance of this in terms of national investment; it is possible to reduce peak domestic demand by 50 % at a cost of less than £50 a kW. Compare this with the cost of installing new energy getting, converting and distribution capacity; nuclear power at £300 per kW; north sea oil at £180 per kW.

People in Houses

So far I have dealt only with the technical problems of the building fabric in a theoretical way, but, while the houses can be readily classified, the people in them cannot. Half of them are tenants, the rest owner occupiers; their incomes, and what they think worth spending their money on, vary widely, and so do their heating needs and preferences.

Consider standards of heating. A substantial number of our dwellings are practically unheated, others are still by most standards overheated, and several million houses with whole or partial central heating have no insulation. There can be no doubt that at present, for many households, thermal insulation would simply mean more comfort and no reduction in energy consumption. For some—in particular the increasing proportion of the aged—improved insulation could mean that more of them will survive cold spells. For others—those with virtually no heating, or inefficient heating, such as unmodified open fires, many millions of which are still used—insulation would provide neither extra warmth nor reduction in energy consumption, simply the possibility that if and when the occupants are able to afford some proper heating there is a chance they may be able to get some benefit from it.

Or consider the fact that half our householders are tenants. Where the owner also pays the fuel bills there is at least a chance that market forces—especially the so called policy of realistic energy pricing—together with education and exhortation may eventually persuade a substantial proportion of owner occupiers to grasp the nettle and 'save it', though to judge by the sales of mineral fibre insulation, there is little evidence of this happening yet. But there are plenty of owner occupiers who simply cannot find the capital to insulate. And at the same time there are plenty of affluent tenants, as well as some desperately poor ones; the former are neither prepared to invest in somebody else's property nor, doubtless, permitted to do so if they wished, and for poorer tenants it is all a hypothetical question anyway.

It would be naive to imagine that market pressures, which manifestly are operating only sluggishly, even in the owner occupier sector, are likely to have any effect in rented housing. A few enlightened local authorities and a handful of private landlords have recently insulated some of their property, but it is difficult to see the process even continuing, never mind expanding, in the present economic situation.

A factor that further complicates the picture, and one which must have a bearing on any policy for energy conservation in housing, is the variety and wayward distribution within the categories of house and householder, of the many different modes of heating. A few generalisations are possible; starting with the smallest contributor to domestic heating, oil serves the two opposite ends of the market—the well-off, efficiently, in central heating systems, and the poor, damply and dangerously, in paraffin stoves. Electricity, the next in terms of nett energy supplied, but the biggest of all in terms of primary energy consumed, performs an oddly assorted range of roles. At one end of the spectrum, in a handful of highly sophisticated applications allied to ultra high insulation and sensitive controls, performing fairly efficiently, and at the other end, the millions of one and two kW 'fires', frequently the only source of heat for the poorest of our families and old people, but desperately expensive to switch on. Between these two, a variety of off peak applications—block heaters bought on the slogan 'start central heating for £50', underfloor heating, so often the fruit of a deal between the local authority and the area board, with house service charges as one of the counters. In general the most expensive to run, serving the people least able to pay.

Gas accounts for the next biggest share, and, growing fast at a rate of 13% per annum between 1965 and 1972, will soon be the biggest. It goes predominantly into central heating applications, in a mixture of appliances. Over the whole range, central heating, and room appliances, it operates generally in a utilisation efficiency range of 60 to 70%. I hope it is not too much to assume that, with the completion of £500 million worth of natural gas conversion, most of these appliances are working somewhere near their best levels of efficiency.

Solid fuel, which between 1960 and 1972 diminished as a source of energy to housing by almost 50%, is still the biggest single fuel. Less than 20% goes into central heating systems and is used at an efficiency in the order of 60%. The rest—over 80%— is burned in open fires, at probably less than 30% efficiency. Another case of the poorest sections of the population being served by the least efficient heating. Because they are poor they have not the capital—or the collateral—to install the more efficient systems, so they pay more for their heating, which helps to keep them poor which...

Finally, the matter of changing standards, which further complicates the domestic energy scene. About 8 million of the existing stock of dwellings now has some form of whole of partial central heating—7 million of these systems installed since 1960. In most cases, though by no means all, installation of central heating has been accompanied by some improvement in insulation—usually limited to 25 or 50 mm of mineral fibre in the roof space and the lagging of the hot water cylinder, and in most cases no doubt there has been a substantial improvement in the efficiency of the installation itself.

It is largely due to these installations and the little insulation that accompanied them—of which something like 4–5 million have gone into existing houses—that during this period, 1960–1972, the nett energy supplied to households remained constant, despite an 18% increase in the number of households in the same period. I think this is highly significant, for there can be no doubt that most of us have greatly improved our standards of thermal comfort in this period; there are more of us, but it has been done largely in an existing housing stock that previously had no insulation, and all this without increasing energy consumption. Doubtless some people's declining standards—especially the old—contributed to this phenomenon, but the main overall trend cannot be denied; we got more for less, through a myriad of small decisions and investments on the part of millions of householders, diffuse and varied, responding to major central decisions and investments on the part of the energy corporations and, to a lesser extent, equipment manufacturers. Remember Shell's Mrs. 1970 who used to come down to pay the milkman in her nightie? She probably sold more gas central heating than oil, but it was all much the same in those days. What was not so much the same is that at the same time as householders were significantly improving their thermal efficiency and achieving so much more in both quality and quantity without any increase in nett energy consumption, the gross energy input—that is the primary fuel into the conversion and distribution systems serving these householders—increased by some 27%. This discrepancy is almost entirely attributable to the increase in the use of electricity for heating in this period. In national terms, the overall efficiency of off peak storage heating—say about 28%—compares somewhat unfavourably with the performance of the open fire—about 30%—that it so often replaced.

Proposals

The proposals I wish to make are not intended as a policy for implementation, but as a means of provoking the radical thought that this combined problem and opportunity needs. What I have to suggest would need to be part of an overall energy policy designed to achieve a reasonably smooth transition from dependence on 'capital' energy sources—fossil fuels, nuclear energy—to 'income' energy—solar, wind, tidal—over the next century. My proposals are largely time-buying to enable us to eke out present workable reserves of capital energy over the longest possible period. They would release manpower and material resources from large scale centralised capital energy enterprises—nuclear power, north sea oil—for the development and production of the small scale dispersed equipment and the design techniques we need in order to exploit the abundant ambient energy available to us.

I believe we must use both the stick and the carrot.

(1) Domestic energy tariffs should be turned upside down—present 'promotional' tariffs encourage consumption—to provide a small allowance of energy at a cheap rate, and a very steeply rising rate for consumption above the allowance.

(2) The cheap energy allowance should be sufficient to provide a reasonable standard of heating in a properly insulated and draughtproof house of average size, with an efficient heating system.
(3) There would be grants, up to 100%, available to all householders to carry out insulation and draughtproofing, and for the replacement of open fires and electric heating with efficient gas and solid fuel burning appliances.

The cost would be enormous; over a period of ten years or so, it might cost about £500 million a year, but this is in the same order of magnitude as the Central Electricity Generating Board's programme of investment in new generating capacity. But with the drop in domestic consumption I am proposing, this investment would no longer be needed. I am suggesting a massive shift in investment away from energy getting and converting, and I believe that in capital cost alone, it would show a credit balance. The savings in energy—of the order of 10% of total national consumption, or a money value of about £60 million a year, would be a free bonus, going on, without further expenditure, for as long as the housing stock lasts. We can choose to go for growth in the getting and conversion of energy, in which case no doubt the prophecy will be self-fulfilling as 'demand' expands to meet the capacity, or we can decide on a totally different course and start by asking what it is we really want—more warmth for more of our people, or merely more energy going to waste.

REFERENCES

1. *Energy Conservation in the United Kingdom*, NEDO, 1974.
2. *Energy—Waste Not, Want Not*, Tyneside Environmental Concern.

Discussion

Dr. S. Leach (*Building Research Establishment*): Ted Nicklin in presenting his paper asks specifically if the Building Research Establishment's economic analysis took account of the capital cost per kilowatt of generating electricity. The answer is yes indeed it does. Long run marginal costs include full capital costs, that is the capital and interest payments in new plant. On the other point that he raised we are not frightened by the £500 million per annum he mentioned or for that matter the £10 000 million I mentioned in my paper which is the order of magnitude of investment needed for the 15% saving of the national primary energy. It is indeed comparable with the investment in North Sea oil which is currently being spent. At the Building Research Establishment we are providing information about energy use, about the potential for energy conservation and the costs and savings involved so that the right action can be taken. I was pointing out the consequences of trying to spend a sum of this size quickly.

Karen Squire (*Mrs*) (*Esso Petroleum Company Limited*)*:* Mr. Nicklin mentioned the growth in penetration of central heating: I have been looking at this and it appears to follow to some extent the traditional pattern of the saturation curve for domestic appliances. There has been continuous growth, as he pointed out, in the last 10–15 years in the number of systems installed, and since it has apparently made no response in the past to economic factors, I think it adds urgency to his case for increased insulation in domestic premises that we expect a growing number of large-volume domestic users in the country. I wonder whether he is prepared to do any 'crystal gazing' as to the future rate of installation of central heating systems, and whether he expects it to be as strong in the next few years as it has been in the past.

Secondly, I agree that a system of promotional tariffs in the present situation is deplorable, and while I find his proposals very thought-provoking, I am a little suspicious of his word 'cheap'. I would like to know whether there is some written economic analysis behind the proposal (for turning tariffs upside-down) that I could refer to. A lot of the costs of supplying energy are fixed, and it is doubtful whether they would fall a great deal if the volume of consumption per user dropped—therefore, I do not really see that the word 'cheap' is applicable.

F. E. Nicklin: On the last point I did say relatively cheap and I did not specify what I mean by cheap. I am talking about a small allowance of energy at a much lower cost than the next lot. There would have to be some rapidly rising scale for energy which is over and above the basic allowance. I certainly am not going to do any 'crystal gazing' about the rate of penetration of the market for central heating—the future is very uncertain indeed. I am quite certain however that expectations on the part of a very large proportion of our population have been excited and have been created by the advertising that has gone on and by a whole range of things, by television, by any number of powerful agencies etc. If expectations are defeated by increasing energy costs then we shall have a very powerful ground swell of opinion throughout the country which will perhaps help to reinforce the demand for a rational energy policy. I see no reason at all why the expectations of reasonable comfort in most of our houses should not be fulfilled.

H. M. Roos (*Gasunie, the Transport and Selling Company of Natural Gas in the Netherlands*)*:* I should like to ask Mr. Nicklin about cavity insulation with Urea Formaldehyde. What are your experiences in Britain concerning the life of the Urea Formaldehyde foam and what about transfer through the walls.

F. E. Nicklin: I think perhaps The Building Research Establishment (BRE) are better placed to answer this question; I have not heard of problems arising from the foam in cavities. Certainly one has heard a good deal about the problems of inadequate workmanship in putting this into walls and BRE have produced recommendations on the situations where it is not recommended to use it, or at least where very careful consideration has to be given before it is done.

Dr. S. Leach: There is a lot going on at BRE in the Materials Division looking at cavity walls which have been filled for a long time using Urea Formaldehyde foam to see if it does in fact retain its properties. It has been found to show evidence of cracking. In certain situations there has been a transfer of moisture across the cavity and this is why as Mr. Nicklin said the guidance we are giving is that cavity fill is more appropriate in certain situations than others. Where there is a high incidence of driving rain this obviously increases the moisture transfer across the cavity. However, this is not necessarily the end of the world when it happens. There are remedial treatments being developed which would take care of the situations where moisture transfer had become a problem after the 'fill' had taken place.

P. R. Achenbach (*National Bureau of Standards USA*)*:* I infer from a number of statements that Urea Formaldehyde foam is used extensively to fill cavity walls. Dr. Leach spoke about the later transfer of moisture—I wanted to ask you about the dissipation of the initial moisture. Is there any problem with blistering of paint and so forth due to the initial dissipation of the moisture?

Dr. S. Leach: I have not heard of any problem such as blistering of paint. In the materials used in UK housing, around the 'fill', moisture transfer exists so it is possible for the moisture in the foam to diffuse out. There have been measurements made of the properties after a period of time and it is true to say that the useful effects of the cavity 'fill' start to become apparent very quickly. There is not a major problem of moisture from the foam itself.

Dr. P. V. L. Barrett (*ICI Insulation Service Limited*)*:* I would like to respond to the questions by Messrs. Roos and Achenbach.

To answer the three points raised by Mr Roos. There is no real evidence that conditions within the cavity cause significant deterioration of Urea Formaldehyde over the life of the building. As with all plastics, one can conduct accelerated aging tests but these require conditions which are not representative of those within the cavity. Secondly, cavity 'fill' is not just any old job. To set up the Dutch operation of ICI it took us 6 weeks to train the initial operators. This gives a measure of the degree of skill required. Performance in use was the third point. I can say that last year, which was an incredibly wet year, when insulating local authority houses over 3 years old, where water penetration resulting from constructional problems had been resolved, our own experience here was of 4 cases in 10 000 of water penetration. Each was subsequently diagnosed and rectified. Now this shows,

I would submit, that used correctly, Urea Formaldehyde foams can be a very safe material in use. Performance figures on the incidents of water penetration in new constructions have been produced by the Building Research Establishment and others where cavity 'fill' has been used. The average was found to be 2·4%, our own figure is 1·2%. These figures are substantially below the 3·4% to 5·8% incidents of water penetration found by the National House Builders Regulation Council for new constructions without cavity 'fill'.

Finally, Mr. Achenbach's point—the amount of water that is added to the building structure by cavity fill because it is a wet process is very low indeed. Dry foam is only 0·5% by mass and the additional process water that is associated with the wet foam only adds 1 to 1·5% over the thickness of the wall—well within the capacity of the brickwork.

J. M. Barber (*University of Liverpool*) *wrote:* How can the energy tariffs be altered to provide a small initial allowance at a cheap rate with further supplies at a steeply rising rate, without the following happening: (1) The consumer purchases his initial cheap allowance of, say, gas and electricity. (2) The consumer then makes up his energy deficiency by buying 'batch purchase' fuels such as solid fuel, oil, L.P.G. etc. which are readily obtainable from a multitude of sources. Therefore, unless there is a very fundamental re-organisation of the energy supply systems such that all energy is obtained from only one supplier (*i.e.* a complete energy monopoly), I suggest Mr. Nicklin's suggestion is unrealistic.

F. E. Nicklin, (*wrote*). In order to benefit from paying the tariffs in the way Mr. Barber suggests, a consumer would need a diversity of heating systems within one premises. I would not imagine that this is a common enough situation to worry about, and it is hard to visualise a domestic consumer spending money on new installations to take advantage of more than one cheap tariff, when all he has to do is to insulate, and otherwise thermally improve, his house. Certainly a complete energy monopoly is not contemplated.

11

Energy Saving in Industrial Buildings

E. J. ANTHONY and J. W. HERBERT

Presented by R. B. WOODCOCK

Synopsis

Looking at the Manufacturing Industry there appear to be three phases whereby energy saving and conservation can be practised:

Phase 1
General housekeeping measures incurring zero capital expenditure, making people aware basically by publicity that energy can be managed by everybody no matter what their position or level in the organisation.

Phase 2
Measures that can incur capital expenditure—but where that expenditure will be minimal. Such measures can include the revision of standards of heating, lighting, ventilation, etc. Where changes can be made to manufacturing processes that will thereby conserve energy then these should be put into effect.

Phase 3
Measures involving high capital expenditure either in process changes, building or building services changes, or in the utilisation of process by-products.

Examples of specific measures taken in each of the categories will be given as well as a further range of possible examples. Two main themes will run through each phase, namely the participation with all levels of employees and the full use that must be made of publicity. The paper will be balanced by co-authorship; Management from a large manufacturing industry with energy cost approaching £2 million per year, and a Consulting Engineer practising in the industrial buildings field.

Other papers being presented during this Conference discuss and define the techniques available to conserve energy. This paper is intended to illustrate the problems which face a large manufacturing company when a sharp change occurs in the availability and cost of the fuels which must be used to enable production to continue and to say what was done and is being done in these circumstances.

When asked to present this paper the terms of reference required that dimensions should be in S.I. units. This gave an immediate problem for the units of measurement mainly being used are Pounds/Pence which are not within the S.I. range.

The salient point to be made is that energy conservation in the context of a commercial enterprise is subject to exactly the same financial evaluation and justification as any other investment. In the present climate one may wish to install energy conservation measures purely to conserve resource without economic substantiation. Any energy saving project which does not meet the return on financial investment required cannot be accepted.

The Concise Oxford Dictionary defines 'fuel' as 'material for fires, something that feeds or inflames passion.' Indeed it does. Engineers may not like the definition but it has current significance. The emotive pressure which exists must be removed from the subject. It would seem that if it is possible for fuels to be priced at a true cost level then commercial interest and resource conservation interest must be synonymous, assessment of projects would be simplified and the conscience of many engineers would be less racked.

At Perkins Engines Company specific energy conservation really started in October 1973. Previous to that the Company had undertaken cost reduction exercises which included consumption of electricity and heating fuel. The result of these exercises was a disciplined usage of fuel in each of these applications, which led to a smaller usage than one would normally expect in an industry such as the manufacture of high speed diesel engines.

This was the background of the scene set when the energy crisis hit the industry in October 1973. That is, a disciplined standard of housekeeping in the usage of energy. In October 1973 severe restrictions in the supply of fuel oils occurred. This was rapidly followed by an escalation of prices the like of which had never been seen in this country.

The situation which then faced Perkins Engines Company was perhaps almost unique in that higher fuel prices could be expected to stimulate the demand for diesel engines in that the high speed diesel is some 30% more efficient in its use of fuel than that of a gasoline engine. It was highly likely that fuel oil would be controlled and as allocations could be expected to be based on the previous year's usage it seemed that Perkins might suffer more than most, simply because the company had already economised. Increases in price would considerably increase the operating cost and have an adverse effect on profitability. The Company realised that under these conditions an energy conservation campaign was required that went much further than the cost reduction exercises that had been undertaken in the past.

At this time Perkins' planned usage of the various grades of fuel oil was 6 million gal per year, that is, about 20 gal for every engine built. The planned usage of this fuel could be broadly split by volume as follows:

Heating of factories and offices	3 300 000 gal	54 %
Production engine testing	1 240 000 gal	20 %
Engineering research and development testing	1 054 000 gal	17 %
External transport	434 000 gal	7 %
Internal transport	172 000 gal	2 %

In addition to this, Perkins planned to use some 60 million units of electricity and small quantities of coal and gas. At mid 1975 prices the energy bill stands at approximately £2 million per annum.

The principal energy entries to the complex comprise three boiler houses having 16 boilers of 67 000 kW (steam) total capacity. Steam is generated at 552 kN/m^2 and reduced in many instances to 173 kN/m^2 for use. Electricity comes in via a single intake substation of 24 MVA capacity at 11 kV.

The heating of factories and offices being much the greatest user of fuel this was obviously the area in which to take effective action. This energy is largely used in space heating and since most of the buildings are of a permanent construction having U values better than average factory construction, the fabric loss of the buildings could not easily be improved by better insulation. The most obvious areas in which any economy could be effected were in the air change/infiltration loss and in the adoption of reduced heating standards.

Phase 1—*Good Housekeeping*

At an early stage in the consideration it became obvious that the immediate contribution to effective savings was going to rest with the full co-operation of all employees. In order to achieve this the specific participation of higher management had to be made apparent to all staff and all employees had to be made aware of the problems and the actions required of them in order that their co-operation could be effective.

The first action taken was to appoint a Senior Director as 'Fuel Ombudsman'. It is interesting to note that this sort of effective recognition of the importance of energy conservation was advocated by the Secretary of State for Energy some fifteen months later. The effect of this appointment was that the project became top priority and received the necessary support from all levels of management.

How to communicate effectively to 8500 employees was another problem. With information supplied to them by the Manufacturing Engineering Division the Public Relations Division arranged regular feature articles in the house journal devoted to energy conservation, and prepared and issued to all staff a pocket booklet entitled 'A Guide to Fuel Economy.' The form and content of the guide were designed to bring energy conservation to the layman's level without under-rating the intelligence of the reader. To stimulate all round energy economy a section of the guide was entitled 'Save Fuel in Your Home and Car.' 'Save Fuel' posters were placed throughout the factories and offices together with stickers urging everyone to 'Use Less' and 'Close It.' Posters and the guide had a common theme of 'Beat The Shortage,' the posters being designed to give impact to the theme and advertise the guide which defined the everyday actions which everyone could take to 'Beat The Shortage'.

The Company had been operating a Suggestion Scheme for a number of years. This gave awards of 25 % of the first year's net saving on an idea accepted up to a maximum of £1000. The fuel crisis placed emphasis on fuel saving ideas and therefore in early 1975 it was decided that suggestions relating to energy conservation should be rewarded by 50 % of the first year's net saving on ideas accepted, up to a maximum of £2000. Suggestion forms were included in the pocket booklet in addition to the normal availability of forms at post boxes distributed throughout the works. This set the scene for an energy saving campaign that was to prove both demanding and rewarding.

The guide urged everyone to keep doors and windows closed and to keep ventilation down to a tolerable level. The response was good and resulted in a significant reduction in space heating requirements, which in turn resulted in the using of less fuel. Emergency steps were taken to reduce draughts—in many cases these merely consisted of masking the offending area until effective permanent draught proofing could take place. The application of such measures may seem to indicate a poor state of building maintenance but this is not so. Even good factories, and, to a lesser degree, works offices, have extraneous infiltration outside the ken of the design manuals and this deficiency is normally overcome in reality by margins which really constitute overheating. Much further practical study in this area should be rewarding to designers and ultimately to manufacturers.

Most of the suggestions made related to conditions which existed local to individual work stations, such as casual access doors which needed closers, changing the ratio of tungsten lamp wattage to mercury vapour lamp wattage in mixed lighting fittings, etc. No awards of £2000 have yet been made but several of over £500 have been awarded. Action on suggestions accounted for approximately 15 % of the savings achieved in the first year. It soon became apparent that the combination of all these steps would not give the reduction in heating fuel oil usage which had been targetted at 20 %. More drastic action was needed.

Clearly, in order to achieve target the temperature to which the factories and offices were heated would have to be reduced. Full discussion with the Trade Unions resulted in an agreement to reduce the space heating temperature by approximately 3 °, to 20 °C in offices and 17·5 °C in workshops. The result of these economies in heating fuel oil was that during the first year Perkins used 26·4 % less heating fuel than was planned. In cash terms the saving in heating fuel was £76 000. In volume, 910 000 gal. This is equivalent to 1·86 gal for every minute of every day of the year.

Two other aspects of the heating systems were examined, namely the generation and the distribution of the heat used for space heating. In the generation of heat, the main saving that could be made was to ensure that the boiler combustion was as efficient as possible. It was also necessary to regularly monitor the results being achieved to ensure the combustion efficiency was being maintained. With this in mind, Perkins arranged for a fuel technologist from the oil company who supply the fuel to inspect the boiler houses and recommend the best method of obtaining optimum benefit from the fuel used. Burner adjustment resulted in a 2–3 % increase in efficiency, which the training of the foreman who attended several courses appears to have maintained. It is interesting to note that the oil company concerned is now using the Perkins example to illustrate how they can help other companies to save fuel.

With heat distribution a first aid exercise was put in hand. This gave priority to repairing leaks in the system and replacing broken lagging. Other general checks were made to indicate where immediate savings could be made or where possible measures to reduce energy existed which could be put into a programme for future action, these being:

(a) Check air volume of all fans and if in excess of requirement reduce by changing fans and/or motors if viable.
(b) Check running loads of all motors against rating and if not correctly rated for present usage corrective measures taken.
(c) Check all set points and ensure systems are working as intended.
(d) Ensure that the preventative maintenance system includes checks of controls and of the efficacy of continuous energy monitoring equipment.

Phase 2—Minor Capital Expenditure Works

Let us now look briefly at the second phase, or the medium term measures. Typical examples of the type of work that is being carried out are improving roof insulation, particularly in offices, improving pipe insulation paying particular attention to flanges and expansion joints, improving space heating control—here additional programmes have been applied to ensure that the heating is only on at the time it is required and thermostats have been fitted to ensure that over-heating is avoided. Plastic strip door curtains are being fitted in certain places to avoid the ingress of draughts and to eliminate the need for hot air curtains in these situations.

Experiments are being carried out with pre-heating combustion air for the boilers. This involves fitting trunking from the air intake so that the air for the boilers is picked up from a close vicinity to the chimney stack enabling some of the heat which would otherwise be lost from the stack to be recuperated and used as input air for the boilers. This is currently in the experimental stages and it is too early to say how practical it will be. However, indications are that it may reduce the heating fuel required by something like 4%. Since this measure is only likely to cost approximately £200 in direct costs for each boiler it is well worth trying on one boiler.

Plans are also in hand to interlock the factory space heating in selected areas with the factory ventilation. This will involve wiring the heating in such a manner that when the ventilation is operating the heating is not. The effect of this will be that the ventilation can be used for a short period to remove any fumes from the factory but the wasteful lengthy use of the roof ventilation will be discouraged as it will cause the factory temperature to drop significantly. This work is unlikely to be completed in time to be effective for the 1975/76 heating season but Perkins should have full advantage of it for the following season.

The heavy fuel oil used for space heating has to be stored at a temperature of approximately 40 °C. This means that the heat loss from the storage tanks is quite significant. Arrangements are in hand for these tanks to be insulated which will reduce the heat loss from them by approximately 75 %.

These are examples of medium term projects which are being undertaken. Some items that were not economically viable two years ago now give a very rapid return on capital. This particularly applies to improved insulation; in fact the heavy fuel oil tanks had been the subject of a cost exercise at an earlier date but the measure had not been viable at that time when fuel was so much lower in cost.

Many matters are being given priority but it will be some considerable time before all the projects are fully completed. In the case of insulation within the factories times of access are very limited and work can only be carried out when production is not taking place, which in some cases is only during the holidays.

Experience is showing that medium term investments in the energy conservation field give a very satisfactory return on capital, generally between 3 and 18 months. The savings effected in the first phase are being used to finance the second phase programme.

SAVINGS IN ELECTRICITY

The other major aspect, namely electricity consumption, was approached in a similar manner with a first and second phase strategy. The communication campaign again proved invaluable. Good housekeeping was a most important part of the first phase. It stressed that action was needed by all and gave a number of recommendations of the type of action that could be taken. These included, switching off all equipment when not in use, switching off lights when natural light was adequate, avoiding the waste of compressed air (this included the repairing of leaks in the compressed air distribution system), and ensuring that nothing electrical was left on when the premises were not occupied, unless of course it was required for safety during this time.

Lighting standards were reviewed and in areas not associated with a high degree of work activity such as storage areas and circulation spaces, levels of illumination below the IES Code have been adopted but care has been taken not to introduce glare problems, particularly when higher intensity illumination was required for isolated local work functions. At this point it must be stressed that in some assembly areas levels above the IES Code are still considered necessary and are being retained. Ineffective general shop lighting such as that positioned over solid roofed shop floor offices, etc., was removed.

On the tariff selection aspect the Company evaluated the cost on all published tariffs available to them and compared each of these with the existing tariff. None showed any lower cost. The Company then examined its situation to see if there were any extenuating circumstances to justify requesting a special tariff. A significant point which arose from this was that 32% of the electrical units used were in fact used between 2300 and 0700 hours the following morning. This period is regarded as off peak by the Electricity Authority. The Company therefore approached the local Electricity Board requesting that a special tariff be devised taking this fact into consideration. The Electricity Board agreed with this and gave a special tariff which gave reduced unit charges during off-peak hours. At that time this gave a projected annual saving of £5000. It is interesting to note that when the electricity tariff was revised in August 1974 to make off-peak electricity less attractive, the projected savings to Perkins were reduced from £5000 to £2500 per annum. However, when the electricity tariff was again revised in April 1975 to again encourage the use of off-peak electricity, the projected savings by being on the special arrangement were in fact increased to £13 000 per annum. This strongly emphasises the need for keeping the tariffs under constant review.

Peak load shedding was the subject of an earlier cost reduction project. Here it was necessary to ensure that the actions initiated in the earlier campaign were still effective. This involved making sure that electricity was not used for storage purposes (such as battery charging and water heating) during the time of maximum demand. Wherever possible, without affecting production, machinery with a large electrical load, but spare capacity, was not used during the time of peak demand.

A bonus from the previous exercise was a maximum demand warning device which continuously compared the actual maximum demand during the current half hour integrating period with a pre-set target maximum demand. When the actual exceeded target, a siren sounded throughout the factories and offices enabling short term action to be taken to reduce the peak demand during the current integrating period. The Public Relations campaign particularly stressed the need for reducing electrical load when the maximum demand siren sounded, providing, of course, this did not affect production efficiency, quality or safety.

Second phase electrical measures fall into two groups:

(1) Direct energy saving measures, *i.e.* reduction of lighting intensities and optimisation of motor sizing.
(2) Measures which reduce electricity charges. These latter measures, *i.e.* peak lopping and power factor correction, do not save energy in great degree within the manufacturing complex itself, except for the distribution copper loss which is effected by the current reduction due to power factor correction. However, these measures do, if effected throughout the country, have a considerable saving in system losses in that the less efficient peaking plant need not be used so that the mean efficiency of generating plant and the load factor of the system is improved which does represent a very considerable annual energy saving.

The system power factor was found to be running at 0·88. This is a relatively low figure. Investigation showed this was caused by additional machinery being added during the previous year without any additional power factor correction equipment.

It was obvious that an improvement was required. The question was, how far could this improvement be economically taken; the EEB recommended a power factor of 0·98. Perkins find static capacitors the most suitable means of improving the power factor because the processes do not call for any large continuously running motors for which synchronous machinery could be used. A tabulation was made to show the amount of capacitance required to improve the power factor to various levels and the savings potential of each 0·01 improvement in the power factor was calculated. With this information, plus the price of capacitors, the economic viability of improving the power factor to various levels was evaluated. The 'pay-back' period for the improvement in the power factor from 0·97 to 0·98 was only marginally above two years. In fact, the total cost of improving the power factor from 0·88 to 0·98 was £17 515. The annual savings based on electricity costs during the early part of 1974 was £12 110 giving a predicted capital return in 1·45 years. Unfortunately there was no benefit in the first year due to the time needed to obtain and install the capacitors. However, with the price of electricity increasing the savings gained are also increasing.

This campaign resulted in 12·1 % less electricity being used than was planned during the first year. In cash terms this was worth £47 500. For the second year this saving should increase by at least the power factor improvement of £12 500. This alone increases the saving in electricity to 15·3 %.

Phase 3—*Longer Term Measures*

Any manufacturing complex is continuously changing and it is necessary that from time to time the user requirement of areas should be redefined and the environmental conditions necessary for the requirement should be compared to those which actually exist in the area. Inability to establish required conditions is always apparent but the use of oversized plant and systems can be a source of large energy loss. If whole systems are oversized then the possibility exists of saving energy by using the extra system capacity to permit optimum start for the reduced duty running to be adjusted to effect some economy, but it has to be considered whether more saving can be achieved by matching motor power input to the new requirement so preventing unnecessary losses due to running at low load, low efficiency and low power factor.

The adoption of reduced standards requires similar considerations to be given if the maximum advantage is to be taken of such reductions. There is little point in reducing comfort conditions and still wasting energy in constant volume systems which are grossly oversized. Whilst having an awareness of these problems the carrying into effect of the corrective measures requires a continuous effort and cannot be achieved quickly although it may be at relatively low capital expenditure.

In the long term it may be profitable to simulate buildings using a computer model and to check the actual energy consumption with the predicted consumption. It would seem that the arrangement of suitable test and recording procedures so that they can be easily compared to the computer model output data might be an effective tool for energy conservation. Currently, whilst the principle of such systems now seems well established, their use in particular applications requires much investigation and establishment of the variables which exist in industrial buildings. However, until the behaviour of systems is better defined the pursuit of energy economy must be a laborious and to some extent a hit and miss matter. In the long term many aspects have been considered which are likely to involve capital intensive measures, most being waste energy recovery measures.

The breakdown of the total plant annual fuel consumption shows that testing accounts for approximately 2 000 000 gal of the fuel oil used. Although small savings in this area could be expected from care and diligence by the test personnel, the use of fuel under test conditions is inherently highly controlled so that no appreciable saving could be made in this area without a specific investigation of the design and layout of the test facilities. Any economies which can be effected in this area have to be considered on a long term basis and must be subject to close financial scrutiny. Additionally, any measures which are viable have to be planned so that implementation does not cause a lowering of the testing standards or a slowing down in the development programme.

It is apparent that theoretically the heat equivalent of some 1 200 000 gal of diesel fuel is being rejected by the engines in jacket, oil cooler, and exhaust losses and a large proportion of the heat equivalent of the remaining 800 000 gal of diesel fuel is being rejected in the dynamometer brake systems.

Four areas of energy recovery are clearly possible, *i.e.* exhaust boilers, to raise steam or hot water, retrieval of heat at low temperature level from the cell air cooling systems, use of the jacket cooling water in a pre-heater circuit, and, finally, recovery of the energy absorbed in test brakes either by an electrical regenerative method or by hydraulic or pneumatic storage.

There are in this particular case a large number of factors which make the recovery of this large amount of energy difficult and expensive. Firstly, the test facilities comprise some 72 test beds used for endurance and performance testing and 304 production test beds.

These cells are spread out over an area of some 5260 m^2 of factory space. Cells cater for engines of 100 to 350 BHP rating in three different types of cells. Test procedures require engines to be run at various loads between no load and 100 % of full load rating for varying periods. Some cells are run for an eight hour day whilst others run continuously during endurance run periods, any individual bed having a utilisation of approximately 55 %.

An overall investigation of the heat demand profile of the plant shows only limited co-incidence with the possible recovered heat profile, largely because testing takes place throughout the year and the space heating load which is the principal part of the heat demand profile is virtually non-existent for five months of the year. Thus the proliferation of heat reject sources from which the recovery must take place and the load profile characteristics make any centralised recovery scheme non-viable.

This still leaves the possibility of local heat recovery on an individual cell or group of cells basis. Since the maximum temperature rise in cells is required to be restricted to 14·5 °C from air inlet to outlet with the intent of keeping the mean cell temperature to approximately 21 °C as far as outside air temperature permits, large quantities of air are required for cooling under the maximum engine heat dissipation condition with relatively high outside ambient conditions. Since banks of cells are cooled by a constant volume system with heat pick-up shunts, the temperature of the air varies widely with test load. This air might easily be used for heating an adjacent area when such a demand exists by direct injection as supply air. However, the air is considerably contaminated by hot oil fumes and from time to time by exhaust gas leakage. This source of heat recovery could only be utilised by the use of a run around coil, heat wheel or common pipe bank which, whilst being reasonably easily adapted to give transfer from exhaust air to supply air within the same system, are obviously much less viable when the transfer is between systems in different areas because of physical limitations, load co-incidence and control aspects. Thus so far it has not been possible to seriously advocate such methods because of the high capital costs involved.

In the case of recovery of energy from the brake testing equipment, again a number of limiting restrictions were encountered. Where hydraulic dynamometers are used the testing conditions, *i.e.* simulation of marine conditions and the inherent design of existing brakes, mean that the heat is rejected as a very low grade heat which it is difficult to use in the production complex as much of the heat requirement is at steam level and in some cases the heat transportation system used to distribute to air curtains, etc., is a high temperature fluid. Consequently the capital expenditure to utilise heat recovered at low temperature level has not to date been found to be justifiable.

The recovery of heat from the exhaust gas has also been shown to be an expensive operation since the exhaust boiler system cannot be permitted to have a pressure drop which could affect the engine exhaust back pressures on a large number of engines under test. Certain of the test conditions deliberately give a large amount of unburnt carbon in the exhaust on an intermittent basis and to such a degree that a particular gas treatment system needs to be devised to deal with this problem. Firing in to any exhaust boiler to burn unburnt carbon has many problems, not the least being that extra fuel would be used at periods when no heat demand existed. From the foregoing it will be apparent that much work is still necessary before presentation could be made upon which any funding could be recommended.

The possibility of using hydraulic brakes to generate a high pressure water circulation at low temperature to drive a water turbine to generate electrical energy has also been investigated but the development costs predicted have proscribed any further work in this direction.

In cases where electrical brakes are used when running on tests in the generating mode, the generators are used to feed power into the general electrical system. The energy recovered in this way is small and this method introduces operational problems.

Currently the energy developed on test is generally absorbed by a water brake and therefore wasted. The water brake method is the international method used by virtually the whole of the automotive industry throughout the world. It is possible to use this energy to generate electricity. Perkins have recently taken out a patent on a test brake which generates d.c. which can be used as a source for the reforming to a 50 Hertz sinusoidal supply of a quality suitable for feeding into the Company's mains system. Although this is still in the experimental stage results so far look promising and it is likely that all test beds will be changed progressively to the new type and that this will lead to significant reductions in electricity costs.

Perkins are also investigating the heating of a group of washing machines. Currently these are heated by steam and sited a long way from the boiler house. This means that distribution losses are high. While this is acceptable in winter when space heating is needed because distribution heat loss helps heat the premises, in summer the loss is a waste, particularly when electricity must be used for extractor fans to keep the heat down. A solution might be to convert these washing machines to dual fuel heating, that is, steam for winter heating and electricity in the warmer season co-inciding with low maximum demand charges. It is a project that needs further investigation but initial evaluation indicates an annual fuel saving in the region of two or three times the capital cost of conversion. A fringe benefit is that one of the boiler houses could be closed completely during the summer. This will give further savings in operational costs and also an opportunity to undertake the necessary preventive maintenance of the boilers ensuring they are at peak efficiency for the winter.

For the future, energy conservation is being considered much more in the early stages of a building. Higher insulation standards are being demanded than was previously the case. The heating and lighting standards adopted for any new building are being looked at very seriously. Higher standards to give improved worker performance have to be re-evaluated in the knowledge of the new orders of cost of energy, labour and capital. Heating and lighting must, of course, be adequate, but it must also be economic and controllable. The heating of high bay areas in particular must be designed to reduce high level heat losses by the controlled use of internal re-circulation. The use of variable volume systems should also be used where energy savings can be shown to compensate for the higher capital cost generally associated with such systems.

In the energy climate of the last two years, both from the cost and integrity of electricity supply aspects, it is inevitable that self-generation should receive consideration.

The generation of 60 million kWh of electricity per annum could make available in heat recovered in exhaust boilers and jackets of dual fuel engines or similar approximately 60 million kWh in heat.

It is frequently the case that the viability of total energy schemes is jeopardised by interim losses accruing in the servicing of capital which is spent before a revenue is obtained. This is because the electrical and heat load of any manufacturing complex builds up over many years. In the case of an existing complex it should be possible to reduce such a loss to a minimum. Against this advantage there are several counteracting factors. Firstly, the distribution systems as they exist have not been designed to be fed from a central source. Secondly, investment has been made in existing plant and a return has to be obtained on this. It is possible that existing plant could be considered as being in a standby mode and enable plant margin in the main scheme to be minimised without impairing the overall integrity of the systems. There are many aspects to be taken into account but it is certain that such schemes are and will be continuously in mind and engineering and financial evaluation will be pursued so long as it is apparent that it has relevance.

The only general criterion for the ranking of projects in Perkins is the financial merit of each project. This is usually assessed using the discounted cash flow technique with a minimum acceptable target of 30 % after tax return. However, many other factors which can lead to modifications of this target must be taken into account including:

(1) Legal requirements and industrial relations considerations.
(2) The degree of risk in a particular project, *i.e.* a project may be highly desirable if everything works out as planned, but it could be a disaster if one factor was to vary by only a small percentage.
(3) The main strategic priorities of the Company and the state of its markets.

Because of the recent rapid movement in oil prices and the resulting improved position of diesel engines against gasoline, expansion projects are tending to take precedence over cost saving activities for the limited available capital. The current crop of energy conservation projects, because of their very high cost saving relative to the amount of capital involved and also the legal and industrial relations considerations involved, are maintaining a high position in our priorities despite the current emphasis on expansion.

Perkins have not at this point in time used the Department of Energy's 'Energy Saving Loan Scheme for Industry.' With the existing terms it is unlikely that it will be used in the future because:

(1) Only £3 million is available for all British industry. This is a very small sum and would only satisfy the tip of the iceberg. The plant alone for a total energy scheme for Perkins would take 30 % of this allocation.

(2) The administration costs of complying with the Government's requirements are likely to be high and there are fairly severe restrictions on how the money loaned can be used.
(3) The interest charged is around the normal commercial interest rate. It is more convenient, practical and easier to administer, if all borrowing is from one source with the minimum of restriction on how the money can be used.

If the Government seriously want a scheme of this nature to be generally used throughout industry it must:

(1) Make a realistic amount of money available.
(2) Reduce to a minimum the bureaucratic type of form filling and reporting required.
(3) Place less restrictions on how the money is used, and
(4) Make the loan interest free, or at least charge an interest rate significantly below the normal commercial rate.

Summary

The Perkins approach to energy conservation can be summarised as:

Phase 1—Short Term, Mainly Good Housekeeping, Little or No Capital Expenditure. This depends heavily on effective communication which calls upon the Public Relations Division, needing active use of the Suggestion Scheme, and effective management of energy by everyone who uses it.

Phase 2 is the Modification of Existing Equipment. This nearly always involves some capital expenditure but the return is very quick, always under two years. Generally speaking, this is carried out by the specialist groups responsible for the operation of the particular area.

Phase 3—Perkins are only in the process of embarking on this. The extent to which this will go depends upon the future price of fuel and the 'price' of money.

Longer term measures, require the changing of equipment or the changing of techniques. This always involves capital expenditure. The return in this case may be somewhat longer than in Phase 2. Where this phase really comes into its own is in new buildings and new projects. It enables these to be built with energy conservation very seriously in mind right from the beginning.

What did all this achieve at Perkins? In the first year using Phase 1,

£132 000 was saved on heating fuel, electricity and transport fuel. In the second year it is still too early to say what the precise savings will be but they are expected to increase by something like £100 000 during the year.

All this is helping Perkins. It is helping to keep the fuel bill down and therefore indirectly to keep engine prices down. It is helping the country by reducing the amount of fuel used and therefore reducing the amount of fuel that needs to be imported, and it is helping employees because many of the techniques used can in fact be taken home by the employee and used in domestic circumstances. Basically the approach has been one of simplicity, but concentrated and continuous application is needed to achieve results with these simple techniques.

Discussion

P. Waite (*Drayton Controls*): Mr. Herbert in his paper made some very interesting points about Perkins investment in time and space controls at their plant at Peterborough and it is very encouraging to see a company taking these efforts to effect fuel economy. However, I must challenge some of the figures put out particularly regarding the costs of optimum start control and the potential economies. My own company has been involved now in installing several hundred optimum start systems and only under very exceptional circumstances does an installation cost more than £1000 to install. The usual figure is between £500 and £700 and can be as low as £400. They are a worthwhile measure to consider in buildings large enough for the fuel costs saved to recover the investment within a reasonable period of time.

May I also comment on a point made that a simple time switch should recover up to 90% of the potential savings of optimum start controls. In our experience the only way to achieve this is by very frequent adjustment of the settings on the time switch used.

R. B. Woodcock (*Perkins Engines Company*): I am delighted to hear you produce optimum start control equipment at less than £500. Of the quotations we had the average price was £1200. I believe the figure quoted by the Property Services Agency is £1500.

Regarding the adjustment, obviously you have to set it so that the building is properly heated before occupancy starts and yet avoid excess heating. You do not change the setting every day. If you get an exceptionally cold spell, then you do have to adjust it. You do lose a certain amount with a simple time switch compared with optimum start control and I fully agree with you. It is a question of cost effectiveness, and ability to maintain temperatures satisfactorily. The prices you were quoting would certainly alter our calculations considerably, although the end result would probably be the same.

A. K. L. McCrone (*R. A. Lister & Co. Ltd.*): May we ask Mr. Woodcock how Perkins intend to recover high grade energy from the 1300 engines tested each day? If recovery in an electrical form is anticipated, how will the problem of varying frequency of supply be resolved?

R. B. Woodcock: We could use the engines for generating DC electricity which we then invert through a thyristor bridge to AC so, technically, it is not much of a problem at all. Economically, it is marginal. With our present rate of return requirements it is very difficult to justify. There is an awful lot of energy wasted in our test shop as I am sure you realise, I see you come from a similar environment where you must have a similar problem. We do not like to see this wasted. We have some experimental test beds in fact generating electricity by the method mentioned from the otherwise waste energy of the engine on test, but they cost something like £7000 per bed compared with about £1000–£1500 for a conventional water brake. We have to save an awful lot of electricity to justify such an investment and it does not meet our currently expected return on investment. In my opinion, it is an example where some incentive is needed from the Government.

J. C. Knight (*Consultant*): I should like to make a point about optimum start controls to refute what has been said by Mr. Woodcock. The efficacy of the optimum start control was proved by the experience of applying the principles to many London office buildings in the care of the then Ministry of Public Works. This work took place in the late 60s. All of these buildings had clock controls of the arbitrary stop and start system referred to by Mr. Woodcock, nevertheless, the optimum start controls produced savings of at least 30% of the energy previously used.

12

Built Form and Energy Needs

W. P. JONES

Synopsis

A method of calculating the positive, negative and gross energy flows through a building envelope is proposed and explained. It is based upon meteorological statistics of temperature, sunshine and cloud cover for typical months and makes use of the distribution of outside temperature within 2·5 °C bands over any chosen period of the day or night. Newtonian cooling is assumed in assessing the decay of night-time temperatures inside buildings and the results are presented of a computer-based exercise into the energy needs of a typical office block, covering the influences of insulation, orientation, glazing, lighting shape and mass. A Building Energy Index is proposed, in two forms (one to account for electric lighting effects) to compare building performances. It is concluded that the amount of glass is the dominating factor and that buildings should be light-weight rather than heavy-weight. The method can be used to assess the energy performance of any building.

Calculation of Energy Flow Through a Building Envelope

A method has been developed (Appendix 1) for calculating the thermal energy flowing into and out of a building through its envelope, over any period. The method is independent of the system used to maintain conditions within the building and of the internal heat gains, although the energy consumed by electric lights when examining the optimum amount of glazing is considered. On the other hand, the thermal energy allowance for ventilation is confined to that needed to deal with natural infiltration. The use of the method therefore provides a true indication of the thermal performance of a building, within the accuracy of any physical and structural assumptions made, and can give a numerical estimate that covers the major factors of building design, namely, insulation, amount and type of glazing, height and shape, orientation and mass. In these respects different buildings can be compared and the best choice made to conserve energy.

Methods of calculating heat gains and losses for specific, design, outdoor conditions are common knowledge and present little difficulty. Limited methods also exist for estimating seasonal consumptions of fuel and electricity. The former are irrelevant to the annual energy situation, being concerned with plant sizing, and the latter are not a true reflection of building characteristics because plant and system efficiencies intrude.

The technique presented here goes a good deal further than these older methods. It is concerned solely with the thermal performance of the building, being based on temperature distribution statistics for a particular location and incorporating calculations for each hour of the day and night through the year. Account is taken of solar radiation, radiant loss to the sky, cloud cover and the influence of neighbouring buildings. An allowance is made for the thermal inertia of the structure and the natural decay of building temperatures at night is estimated by making the simplifying assumption of Newtonian cooling.

The Method

The essence of the method is that an equation for the combined, climatic, energy flux (equation 1, Appendix 1) through the building envelope is solved, for each hour of the 24 in a statistically typical day for each month of the year. This climatic flux, *C*, is divided into six (or more) elements, covering the four (or more) walls, the roof and natural infiltration. Positive and negative results are summed separately, for each of the six elements because, since they will not occur simultaneously, cancellation is unacceptable.

Energy flow rates through the roof and walls are calculated using sol-air temperatures that differ from those customarily employed for design calculations; this is because they are based on mean daily outside air temperatures and account is taken of the influence of cloud cover for both gains by solar radiation and losses by long wave length radiation to the sky. For simplicity, a simple sinusoidal variation of outside air temperature is assumed throughout the day.

Solar gains through glass are calculated in the usual way making allowances for direct and diffused radiation, the influence of shades, building storage effects etc. However, to render this procedure relevant to energy consumption, an allowance is also made for cloud cover. Energy transmission through glass, other than by short-wave solar radiation, is dealt with by adopting a concept of sol-air temperature for glass and regarding the windows as having zero time lag and a decrement factor of one.

Statistical meteorological data yield mean temperatures which conceal the fact that the air outside may well be at a temperature above the inside temperature for some of the time, and *vice versa.* Thus heat gains and losses can both occur through glass even though the mean daily maximum outside temperature is less than that inside. To cover this reality, temperature distribution data is used. At various sites throughout the country the meteorological statistics can be processed to give the number of hours in each month that the outside dry-bulb temperature lies within successive temperature bands of 2·5 °C, from −7·5 ° to +32·5 °C. Furthermore, the information may be presented for each hour of the 24 in a typical day in a month. The method uses this distribution data and is able to determine, within a chosen period (say the hours of office occupancy), the number of hours in any month when the outside air temperature exceeds that inside and its mean value during this time. The same is done for outside values below room temperature. Thus gains and losses can be established for transmission through glass and for the infiltration element, which also has no time lag or decrement factor and hence must use distribution data.

Using the Method

A model building, described in Appendix 4, is chosen to illustrate the use of the method to compare the thermal performance of various built forms. The procedure has been to choose a building of 12 storeys and assume it is glazed, single and double, to various extents in its two long faces only. Four different, reasonably practical, wall U-values and two roof U-values have been selected. With an occupied period of from 0800 to 1700, suntime, and the supposition that the building has services (which it is unnecessary and irrelevant to define) which maintain a comfortable inside temperature when people are present, the method calculates the positive, negative and gross energy flowing through the building envelope over the year. Aligning the major axis of the building in different directions and permutations on the other variables mentioned permits the relationship between the built form

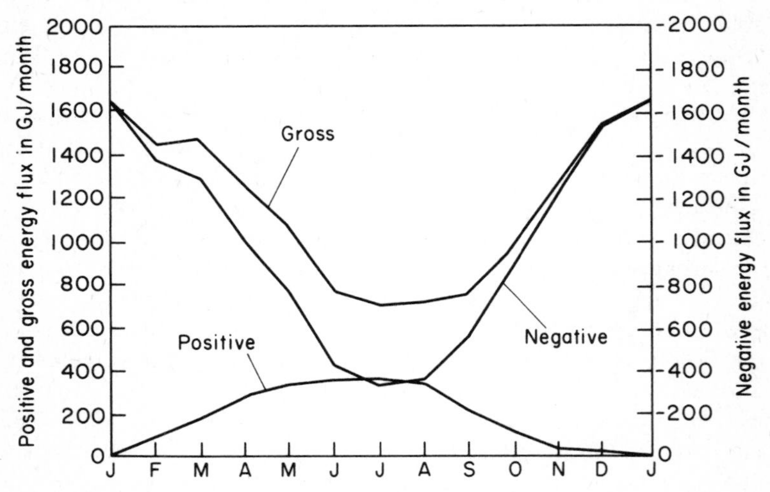

Fig. 1. *The monthly flow of energy through the envelope of the standard, model building, located in London.*

and energy needs to be determined. The quantity of arithmetic is formidable and the method is only usable with a computer.

Figure 1 shows us the energy flux in GJ for each month of the year: positive, negative and gross figures (the sum of the negative and positive absolute values without regard to sign) are shown for a medium-weight, 12-storey building having hollow-pot floors and not very well insulated. Since cancellation cannot be relied upon between positive and negative energies and since a positive energy flow in summer can be nearly as objectionable as a negative energy flow in winter, it is suggested that the gross values are most representative of the energy needs of the building. That is, they show the inadequacy of the building for the admission and release of energy. The best possible building performance would be a straight, horizontal line at zero GJ in the figure. This is obviously unattainable.

A Building Energy Index

To compare the energy needs of various built forms a Building Energy Index (BEI) is suggested. This might be in several guises, expressing relative performance for energy losses, energy gains, or gross energy flows. Further, if we acknowledge that a building must be illuminated, a Lit Building Energy Index (LBEI) is a possibility, to compare buildings with different amounts of glazing.

The BEI compares the energy needs of the built form by choosing a standard building and assigning this a value of 100. In this text the standard is for the 12 storey model building with 50% single clear glazing, $U_w = 1{\cdot}57\ W/m^2\,°C$, $U_R = 1{\cdot}1\ W/m^2\,°C$ and of medium weight (see Appendix 2 for notation). In other words it is a fairly typical office block, not very well insulated according to present views, and with hollow pot floors. Typical energy needs for such a building are as follows:

TABLE 1

	GJ yearly totals			BEI values			
A_g	+	−	*gross*	+	−	*gross*	U_g
0%	49·7	7035	7075	2	62	52	5·6
25%	1181	9225	10406	51	82	76	5·6
50%	2313	11296	13609	100	100	100	5·6
75%	3445	13278	16723	149	118	123	5·6
0%	49·7	7025	7075	2	62	52	2·8
25%	1171	7746	8917	51	69	66	2·8
50%	2283	8593	10876	99	76	80	2·8
75%	3399	9335	12734	147	83	94	2·8

We see that a gross BEI of 100 corresponds to a yearly energy flux of 13 609 GJ and, *pro rata*, it falls to 80, for example, if double glazing is adopted. The various permutations of U value, glazing etc. can now be compared.

The Influence of Wall Insulation

The amount of glazing obviously affects the significance of wall insulation. Thus for the model building, with single glazing, the computed results are:

TABLE 2

U_W	$A_g =$	0	0·25	0·5	0·75
		BEI values			
1·57	gross	52	82	100	123
	−ve	62	76	100	118
0·96	gross	42	69	95	120
	−ve	50	73	94	114
0·42	gross	33	62	90	117
	−ve	40	65	88	110
0·36	gross	32	61	89	116
	−ve	39	64	87	110

We see that a poorly insulated building with 75 % of its long faces glazed needs 23 % more gross energy annually than does the standard case, whereas a very well insulated building with no glass, only needs 32 %.

Figure 2 illustrates this showing there is a base value for energy flows for glass, roof and infiltration when the walls are perfectly insulated ($U_W = 0$). Even in this case the minimum BEI possible is still 26, for no glazing at all. If we take a more common instance of 50 % glazing, the minimum BEI is about 85. The inference is that the amount of glazing is more significant than the U value of the wall.

Roof Insulation and Building Height

Modifying equation 1 (Appendix 3), we can write:

$$F_n = n(F_1 + F_2 + F_3 + F_4) + F_R + nF_I \qquad \ldots 1a$$

for a building of n storeys. The output from the computer can thus be easily processed to illustrate the effect of roof insulation, comparing a well-insulated building (25 % double-glazed and $U_w = 0{\cdot}36\ W/m^2\ ^\circ C$) with a standard case (50 % single glazed, $U_W = 1{\cdot}57\ W/m^2\ ^\circ C$). Figure 3 shows us the results. As expected, we see that it is more important to deal with the roof of a low building than with a high one; a drop of the BEI from 129 to 103 occurs for a single-storey, standard building when we improve U_R from 1·1 to $0{\cdot}2\ W/m^2\ ^\circ C$. About three storeys seems to be the significant height. If

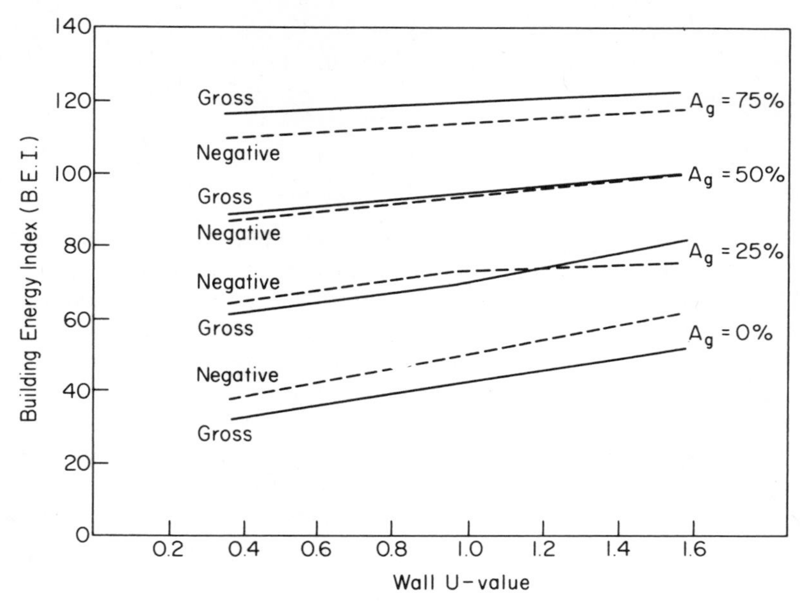

Fig. 2. The effect of wall U value upon the Building Energy Index. The standard building has an Index of 100 *for* $U_w = 1{\cdot}57$ *W/m² °C and* $A_g = 50\%$.

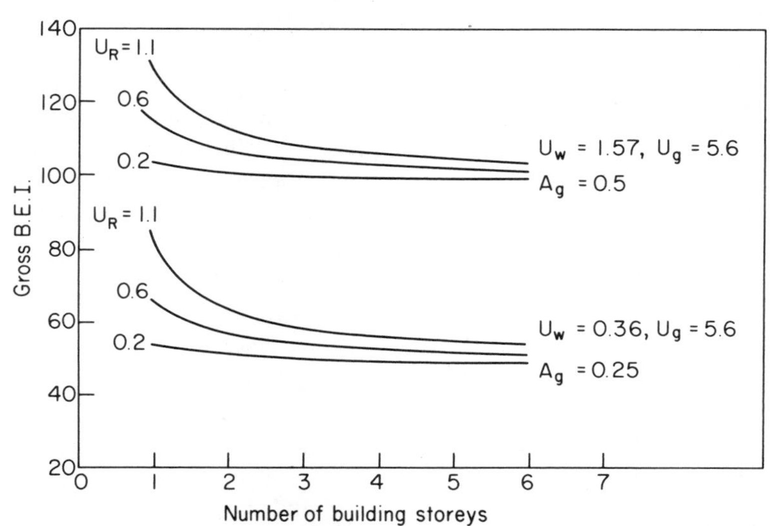

Fig. 3. The significance of insulating the roof slab in a multi-storey building.

we improve the insulation on the rest of the building generally, the picture is similar but the BEI values fall by about 50—clearly a desirable result.

Orientation

There is already evidence[1] that aligning the major building axis E–W gives a lower maximum cooling load than doing otherwise. Calculations of actual energy flow rates over the year now confirm this, as Fig. 4 shows. Two sets of BEI values are plotted: positive ones, corresponding to heat gains and gross figures, corresponding to overall energy needs. Since the heat loss greatly exceeds the heat gain, in this country, and is virtually independent of orientation, the gross BEI values show only a slight preference for an E–W major axis alignment; but the positive values are greatly influenced by both orientation and the amount of glazing, showing a definite preference for E–W. Reducing the amount of glass is very significant.

Single and Double Glazing

Table 1, earlier in the text, showed that double glazing had virtually no influence on the positive flow of energy (heat gains): the positive building energy index improves from 149 to only 147, with 75 % glass, by using double glazing. We do see, however, that the negative index improves to 76 and the gross value to 80. This might tempt us to think that double glazing can be economically justified. This has been looked into many times before and the consensus[1] is that, except for buildings like hospitals, occupied continuously, there is no economic case.

Glazing Amount and Lighting

The results obtained all point in the same direction: use a lot of glass and you use a lot of energy. A building with no glass seems to be the ultimate, but some illumination is necessary and this can be partly natural if windows are used.

Figure 5 is based on work by Owens[2] and shows the illumination achievable naturally by windows with Venetian blinds, the slats being

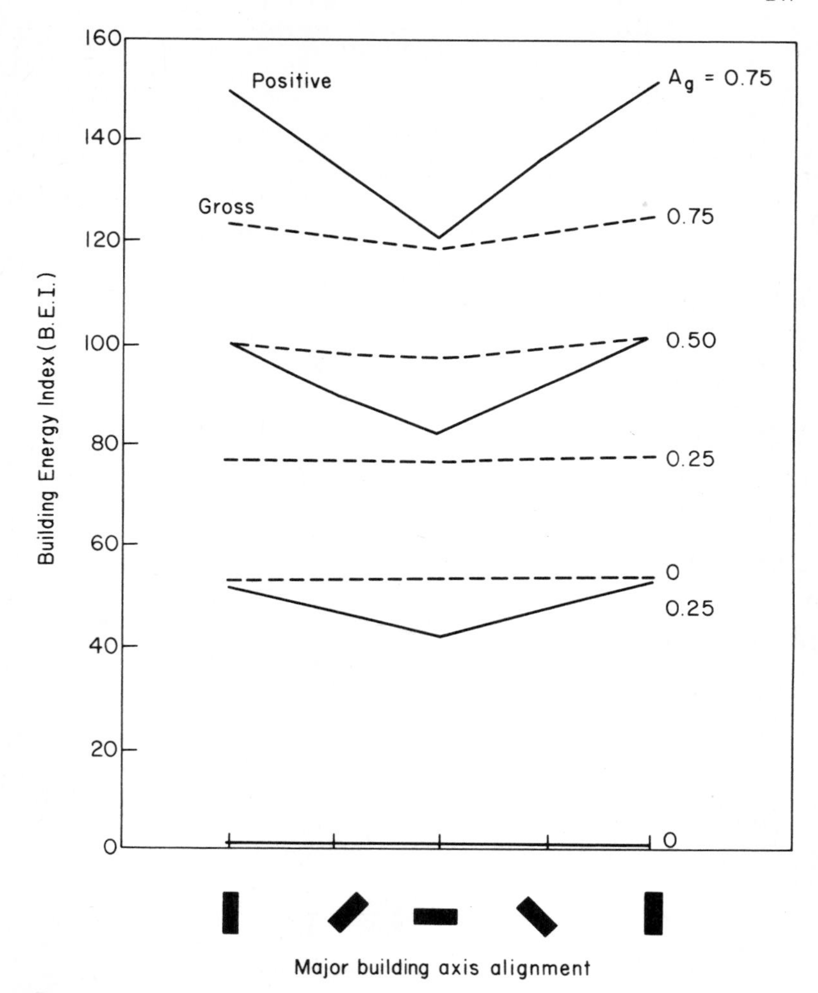

Fig. 4. Orientation and Building Energy Index. Heat gains are influenced by both orientation and the amount of glazing. Heat losses are virtually independent of building alignment, unless large areas of glass are used.

adjusted by the occupants. Daylight outside is taken as 20 000 lx on a flat, horizontal surface in June or July. So 500 lx can be achieved at a distance of about 3·4 m inwards from a 50 % glazed facade.

Making allowances for cloud cover (Table 3), and monthly variations in solar intensity (IHVE Table A6.8), the mean daylight

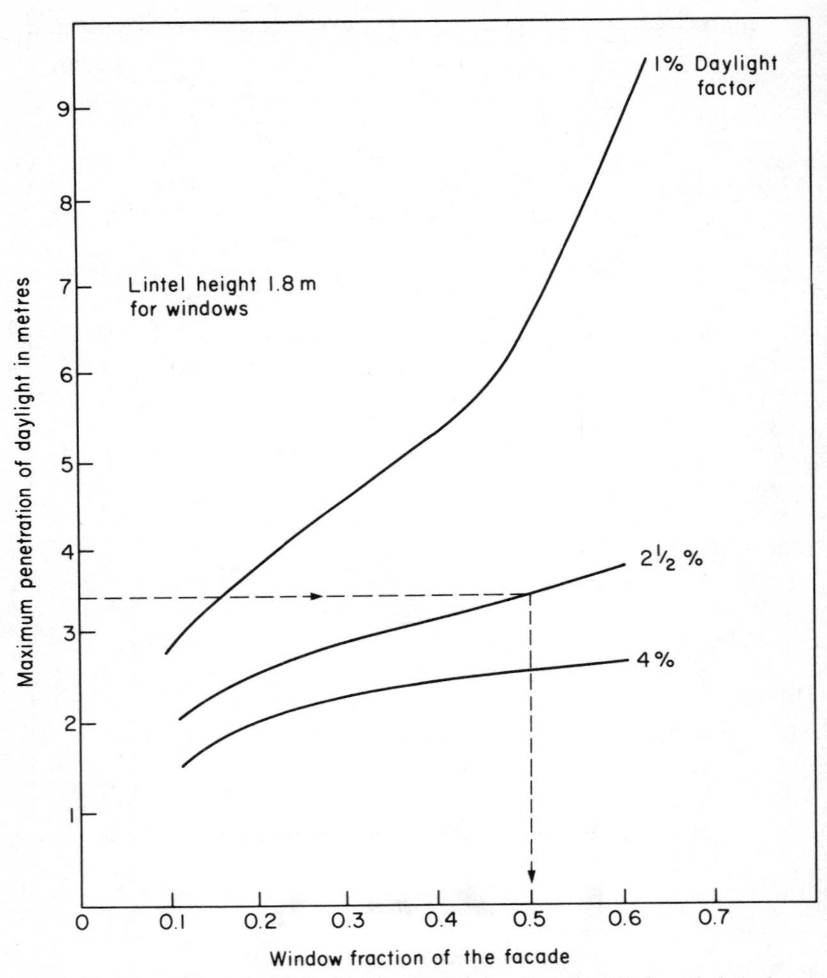

Fig. 5. The penetration of daylight through a window. Venetian blinds, with their slats adjusted by the occupants, are assumed and daylight outside in summer is taken as 20 000 lx on a horizontal surface.

penetration distances were calculated from Fig. 5 to achieve 500 lx naturally, for different amounts of glazing. This gave an area for each month which could be assumed as naturally lit, the remainder of the floor being illuminated by using 25 W/m^2 of electricity, and the following are some of the results obtained, for artificial lighting:

TABLE 3

Month	Jan	Feb	Mar	Apr	May	Jun
	25% glazing on two long faces					
Naturally lit width in m	0·37	0·71	1·23	1·86	2·33	2·45
Mean W/m^2	24·3	23·7	22·7	21·6	20·7	20·5
GJ used	344	307	319	293	291	280

Total annual consumptions were calculated as:

A_g	0	25	50	75
GJ	4140	3731	3623	3470

If we take account of the +ve and −ve indices for the buildings, we can add to them the consequences of the lighting energy. Rather than making a straightforward addition, it is suggested that the 25% (say) efficiency of electrical generation and distribution be taken into account. That is, the energy for artificial lighting should be multiplied by 4, to turn it into a basic energy figure, on the same footing as all the other energy fluxes. A new criterion for comparison is therefore suggested: the Lit Building Energy Index (LBEI), and referred to gross values of building energy fluxes.

The following table compares the results obtained for (a) the standard building and (b) a well-insulated version of it, quoting annual figures for energy.

TABLE 4

	$A_g = 0$		$A_g = 0·25$		$A_g = 0·5$		$A_g = 0·75$	
	+ve	*gross*	*+ve*	*gross*	*+ve*	*gross*	*+ve*	*gross*
Standard building: $U_W = 1·57$, $U_R = 1·1$, $U_g = 5·6$, *Medium weight*								
Building GJ	50	7075	1181	10404	2313	13609	3445	16723
Lights GJ	4140	4140	3731	3731	3623	3623	3470	3470
4 × lights GJ	16560	16560	14924	14924	14492	14492	13880	13880
Total GJ	16610	23635	16105	25328	16805	28101	17325	30603
LBEI	99	84	96	90	100	100	103	109
Well-insulated building: $U_W = 0·36$, $U_R = 0·6$, $U_g = 2·8$, *Medium weight*								
Building GJ	24	4242	1145	6696	2264	9220	3388	11653
Lights GJ	4140	4140	3731	3731	3623	3623	3470	3470
4 × lights GJ	16560	16560	14924	14924	14492	14492	13880	13880
Total GJ	16584	20802	16069	21620	16756	23712	17268	25533
LBEI	99	74	96	77	100	84	103	91

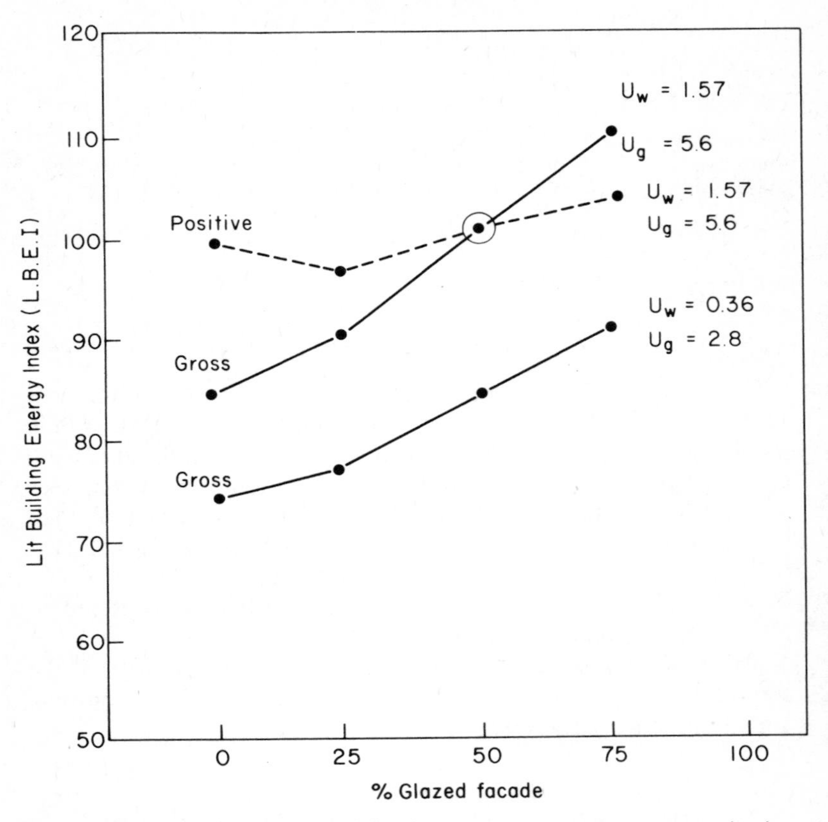

Fig. 6. The influence of the glazed facade on comparative energy consumption in a building lit to 500 lx, partly by natural means but mostly by electricity at 25 W/m².

The above approach is, perhaps, open to question, but it suggests that, even after allowing for the benefit of natural lighting, too much glass wastes energy. If a different standard of lighting had been chosen, say 300 lx, then the picture would change somewhat, but the overall lesson would be the same; the less glass the better.

Psychological tensions are relieved by windows and a building with 25% glass is better in summer than one with no glass, as its positive index shows in Table 4. Figure 6 illustrates these results and suggests that more than about 30% of glass in the outer facade is undesirable, but the attitude to this of people within the building must not be ignored. Much less than 30% may be objectionable to them.

Building Shape

The general principle is that the ratio of envelope surface area to usable floor area should be a minimum, because it is through the envelope that energy flows occur. Mathematically, this leads to a sphere in free space or perhaps a hemi-sphere, on the ground. Practical building restrictions may rule this out but a circular or square plan should be preferred. A further mathematical conclusion is that big buildings are more economical than small ones, because the ratio referred to reduces as the size increases. A previous study[1] of maximum heating and air conditioning loads showed that the optimum heated building shape was a near cube, but the best shape for an air conditioned building had fewer storeys, assuming a square plan and the same floor area. This is because of the solar gain through glass—the more the glass, the lower the optimum building height. For example, a building with a floor area of 25 400 m^2 has an optimum height of 2 storeys when 75 % of its four walls are glazed but 4 storeys if only 25 % are glazed. The two building shapes would be 113 m × 113 m × 6·6 m high and 80 m × 80 m × 13·2 m high, respectively. If the same building has no windows at all the optimum height is 9 storeys and the building dimensions are 53 m × 53 m × 30 m high—nearly a cube.

Those results were based on maximum loads, but there is every reason to suppose that considerations of annual energy flows will yield similar results.

Building Mass

To compare the effects of building mass, three buildings with two standards of insulation and four categories of glazing, have been examined. The standard building is the middle and comprehensively calculated (in this context) yardstick and the extreme instances are buildings of nominally zero and very large mass.

Two aspects of mass must be looked at: first, the mass of the intermediate floor slabs which, in a multi-storey building, is the major influence on thermal inertia and, secondly, the wall mass which is the major factor in energy flux. In all three cases the floor mass has been assumed constant (hollow pots) as far as solar gains through windows are concerned, but a zero mass building has been taken as one with

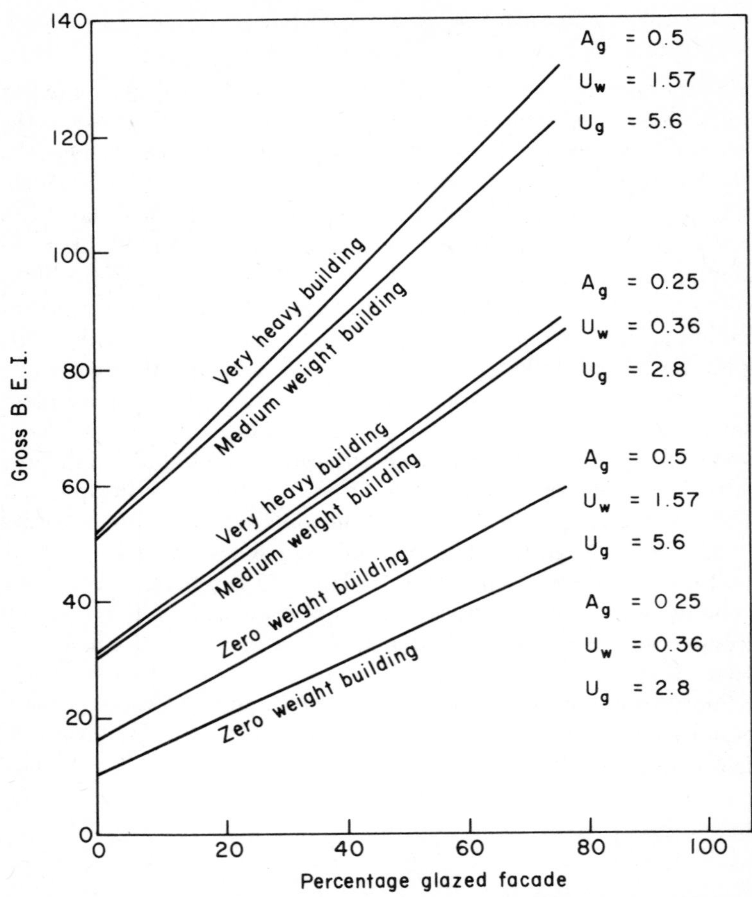

Fig. 7. The effect of building mass on comparative energy consumption. A very heavy building has a wall time lag exceeding 16 *bonus and a time constant in excess of* 500. *A zero weight building has no wall time lag and a time constant of zero.*

absolutely no thermal inertia. That is, at the end of occupancy the room temperature falls to the same value as the outside sol-air temperature and equals it until occupancy starts next morning. Infiltration energy exchanges have also been taken as zero throughout this period but solar gains through windows have been accounted. The building of very large mass has been assumed to have a constant temperature inside through 24 h.

The results are shown in Fig. 7. We see that a large mass building is a bad prospect; buildings should be light-weight in our climate if energy is to be conserved, regardless of the amount of glazing. The argument that a heavyweight building absorbs enough solar heat to offset winter heat losses is fallacious: the climate in this country is too cold. Most of the energy flux, as a comparison of the positive and gross energy flux figures in the earlier text will show, is negative. That is, our problem is offsetting heat loss, not heat gain. A separate, computer-orientated study,[3] has shown that, even for a west wall (without windows) the mean sol-air temperature over the year has a value of only 12 °C, in London, and this is the value about which the temperature inside a building without services will vary, the extent of the variation depending on the mass.

Conclusions

It is concluded that the amount of glazing dominates the energy flux situation and, bearing in mind human preferences, about 25 % to 30 % of the outer facade should be glazed. Double glazing cannot be economically justified for short-term occupancies but may be worthwhile for continuously occupied buildings, U values for walls and roofs should be 0·4 and 0·2 W/m^2 °C, respectively, or better. Major building axes should run E–W if the building is to be air conditioned, particularly if there is a lot of glass. Buildings should be large rather than small and have circular or square plans; heated buildings should be near cubical in shape but air conditioned buildings should be lower, depending on the amount of glazing. Light-weight buildings are definitely better than heavy-weight ones for energy conservation, bearing in mind the well-known problems of condensation with poor ventilation.

The method developed can be used to evaluate the thermal performance of any building.

Acknowledgement

The author wishes to thank Mr. A. G. Pearse for his absolutely invaluable help in programming a very complicated set of equations and data.

Appendix 1—Theory

(See Appendices 2 and 3 for notation and equations)

The climatic energy flow through a building envelope is given by equation 1, which divides the flow into six elemental flux values, four referring to the window-wall facades, one to the roof and one to natural infiltration:

$$C = F_1 + F_2 + F_3 + F_4 + F_R + F_I \quad (1)$$

Each element is defined in terms of hour-to-hour, typical, climatic variations for a chosen month, the dimensions of the building and its physical properties. Thus, for a wall of a given orientation

$$F_1 = zn X_1 Q_1 \quad (2.1)$$

where

$$Q_1 = T_{w1}(1 - A_{g1}) + T_{g1}A_{g1} + S_{g1}A_{g1} \quad (3)$$

and similarly, with appropriate modifications, for the roof and infiltration elements (equations 2.5, 14, 2.6 and 18).

It is assumed that the temperature inside the building is held at a constant value during the hours of occupancy in a given month, but that the night-time temperatures are less. With the exception of T_g, energy flow rates are evaluated for each hour of the day and night, for each typical month of the year. The answers in kW, multiplied by the hours, yield the energy flux for the month in kWh, which may then be converted to GJ, as required, and summed for the year. Positive and negative energy fluxes are separated since it cannot be assumed that they will occur at a place and time which justifies cancellation. Distinct results are worked for the facade orientations, the roof and infiltration.

For the case of transmission through glass, T_g, the approach is similar but, because of the way the meteorological temperature data is handled, the mean energy flow rates over periods longer than an hour are determined. Solar gain through glass, S_g, is treated on an hourly basis.

SOL-AIR TEMPERATURE

A concept of sol-air temperature based on typical days in each month rather than design days is adopted, in accordance with equations 4 and 14, for walls and roofs, to establish T_w and T_R.

A simple sinusoidal variation of outside air dry-bulb temperature (equation 6) is assumed for each particular month, based upon the mean daily maximum temperature and the diurnal range. This is then incorporated into equations 5 and 15, for walls and roofs, to yield their sol-air temperatures.

The equations for sol-air temperature include a bright sunshine factor, f_d,[4] based on meteorological records for the particular location, and items for the heat gain by direct solar radiation, I_t, and for heat loss by long-wave radiation to the sky, I_{LW}. In line with the suggestion in the IHVE Guide,[5] I_{LW} is taken as zero for walls (but not for roofs) when I_t is not zero, it being assumed that long-wave radiation from the ground to the walls in the daytime cancels loss from them to the sky.

Values of I_t are taken from Table A6.8 in the IHVE Guide but I_{LW} is calculated on an hourly basis, or a mean basis for a longer period. The equations for I_{LW} (7, 12 and 16) include an angle factor, B (equation 8), which expresses the fraction of the hemispherical sky seen by the surface;[6] thus for a roof B is 1·0 but only 0·5 for a vertical wall. The presence of adjoining buildings obscuring the sky can lead to a value of less than 0·5 and the assessment of this is left to the judgement of the user of the method. The lower reaches of the atmosphere are not entirely transparent to long-wave radiation, because of their moisture content. Equation 9, defines a factor, K, (Table V), dealing with this, in terms of the mean monthly vapour pressure.[7, 8]

In the case of glass, a sol-air temperature concept is also adopted to evaluate T_g, the energy flux. This is done because glass has a poor U value and no thermal inertia; its surface temperature will therefore rapidly follow the night-time changes of room temperature and energy will be lost to the sky by long-wave radiation in a calculable manner (equations 12) as well as being convected to the outside air. When adopting this approach a reduced U value must be taken for the glass because the radiant component of the outside air film is already accounted for by I_{Lg}.

METEOROLOGICAL DATA

Mean monthly and design values of outside temperature and sunshine are no use for estimating energy fluxes. This is because they seldom occur. Instead, we must use mean daily maximum and minimum values,[9] assuming these to occur at 1500 and 0300 h respectively, with a sinusoidal variation in between.

This is adequate for walls and roofs where thermal inertia smooths out short-term variations about mean values, but it is useless for glass and infiltration, where there is no thermal inertia involved. For such cases an estimate must be made of departures from the mean giving outside temperatures both above and below room temperatures. Their duration at these values must also be assessed. It is possible to do this by processing the temperature distribution data available for various sites in the UK.

Thus for Heathrow airport, in April, data for the ten-year period 1959–1968, inclusive, yields the following within an assumed occupancy period of 0800 to 1700 suntime:

Temperature band °C	−2·5 0	0 2·5	2·5 5	5 7·5	7·5 10	10 12·5	12·5 15	15 17·5	17·5 20	20 22·5	22·5 25
Duration hours	0·5	2·4	11·9	32·4	61·7	74·8	56·2	20·8	7·8	1·4	0·1

If, in April, the inside temperature is 21 °C, then by linear interpolation we can calculate from the above that the outside temperature is above 21 °C for a total of 0·94 h (θ_a) and during this time its mean value is 21·96 °C ($\bar{t}_{aoa}$). Similarly we can determine that it is below 21 °C for 269·06 h (θ_b) at a mean of 10·77 °C ($\bar{t}_{aob}$). The total hours in the month between 0800 and 1700 are 270 (= 0·94 + 269·06). Likewise, the mean values above or below any room value, and their durations, can be worked out for any other period of the day *e.g.* for the unoccupied hours, 1700 to 0800, suntime. Table I lists the values calculated for use in this paper.

NIGHT-TIME BUILDING TEMPERATURES

If Newtonian cooling is assumed and if it is supposed that the heating system starts early enough to bring the building internal surface temperatures up their values of the previous day by the time occupancy starts, then equations can be developed for the cooling-down and heating-up periods in terms of the building time constant.[10,11,12] In fact these curves are often almost straight lines for the range of time constants considered (35 to 450) and so this simplifying assumption is made. To keep the computations within reasonable bounds, inside mean temperatures are taken to be constant during the unoccupied period and some of these are shown in Table II, for different glazing arrangements and heavy and medium-weight buildings, corresponding to 12 storeys of solid concrete intermediate floor slabs and 12 storeys of hollow-pot floors, respectively.

DAYTIME BUILDING TEMPERATURES AND INFILTRATION

A building is assumed to be controlled at a constant temperature throughout the occupied period at the summer design value in July, reducing proportionately to the winter design value in January. Table I lists the values used for the calculation carried out in this text, but any values could be assumed, with an impact, of course, on the inside night-time temperatures and on the processing of the meteorological data.

The same table lists also the infiltration rates assumed. These are taken to be the same at night as in the daytime, but to vary proportionately from 0·5 in July to 1·0 in January.

EXAMPLE

Completing the calculations for a whole year for an entire building is arithmetically very tedious and only really calculable by computer. However, the method may be illustrated by a restricted example. Calculate the energy flux through the east face of the model building (appendix 4) with 50 % single glass in April, if the major building axis lies N–S and internal Venetian blinds are fitted.

1. *Data used:*

Time lag for walls and roof: 8 h; decrement for walls and roof: 0·3; $R_{SOW} = 0{\cdot}05\ m^2\ °C/W$; $R_{sog} = 0{\cdot}05\ m^2\ °C/W$; $f_w = 0{\cdot}8$, $G = 1{\cdot}0$; $U_w = 0{\cdot}962\ W/m^2\ °C$; $U_g = 5{\cdot}6\ W/m^2\ °C$; $U^1{}_g = 5{\cdot}04\ W/m^2\ °C$; medium-weight building; occupancy: 0800–1700 suntime; $E_w = 0{\cdot}9$; $E_g = 0{\cdot}925$.

2. *General solution*

$$\begin{aligned} F &= znXQ \quad W \\ &= 3{\cdot}3 \times 12 \times 86{\cdot}4Q \quad W \\ &= 3421\,(0{\cdot}5\,T_w + 0{\cdot}5\,S_g) + 3421 \times 0{\cdot}5\,T_g \quad W \\ &= F^1 + F_g \quad W \end{aligned} \tag{2.1}$$

3. *Wall flux*

$$\begin{aligned} T_w &= U_w\,(t^1{}_{eo} - \bar{t}_r) + f_w U_w\,(t_{eo} - t^1{}_{eo}) \\ &= 0{\cdot}962\,(t^1{}_{eo} - \bar{t}_r) + 0{\cdot}3 \times 0{\cdot}962\,(t_{eo} - t^1{}_{eo}) \quad W/m^2 \end{aligned} \tag{4}$$

4. *Sol-air temperatures for an east wall*

$$t_{eow} = t_{ao} + R_{so} f_d (\alpha_w I_t + E_w I_{LW})$$
$$= t_{ao} + 0{\cdot}05 \times 0{\cdot}391\,(0{\cdot}8 I_t + 0{\cdot}9 I_{LW}) \qquad (5)$$

$$t_{ao} = t_{15} - \frac{D}{2}\left[1 - \sin\left(\frac{\theta\pi - 9\pi}{12}\right)\right]$$
$$= 13{\cdot}3 - 3{\cdot}9\left[1 - \sin\left(\frac{\theta\pi - 9\pi}{12}\right)\right] \qquad (6)$$

I_t is obtained from Table A6.8, IHVE Guide (1970)

$$I_{LW} = -5{\cdot}77\,BK\left[\frac{273 + t_{ao}}{100}\right]^4$$
$$= -5{\cdot}77 \times 0{\cdot}5 \times 0{\cdot}326\left[\frac{273 + t_{ao}}{100}\right]^4 \quad \text{W/m}^2 \qquad (7)$$

Remember that $I_{LW} = 0$ when $I_t \neq 0$ and *vice versa.* The value for K comes from Table V.

5. *Results of calculating T_w*

t_{eow} is calculated for each hour of the day and the mean sol-air temperature, $t^1{}_{eow}$ is calculated as 11·36°. From Tables I and II we find the mean values of t_r are 19·3° during the period 1700–0800 and 21° during the occupied period, 0800–1700, suntime. Remembering the wall has an 8 hour time lag we calculate the following. All values obtained for T_w are negative and are in W/m², occurring over the hourly periods centred on the times quoted. The symbol θ in the table denotes suntime over 24 h.

Inside θ	*Outside* θ	t_{ao}	t_{eow}	T_w	*Inside* θ	*Outside* θ	t_{ao}	t_{eow}	T_w
0	16	13·17°	14·34°	6·78	12	4	5·63°	4·63°	11·22
1	17	12·78°	13·64°	6·98	13	5	6·02°	6·25°	13·02
2	18	12·16°	12·63°	7·27	14	6	6·64°	12·43°	8·96
3	19	11·35°	10·27°	7·95	15	7	7·45°	16·60°	7·76
4	20	10·41°	9·34°	8·22	16	8	8·39°	18·71°	7·15
5	21	9·4°	8·35°	8·51	17	9	9·4°	19·02°	7·06
6	22	8·39°	7·35°	8·80	18	10	10·41°	18·15°	5·68
7	23	7·45°	6·43°	9·06	19	11	11·35°	16·35°	6·20
8	0	6·64°	5·63°	10·93	20	12	12·16°	14·04°	6·86
9	1	6·02°	5·02°	11·10	21	13	12·78°	14·58°	6·71
10	2	5·63°	4·63°	11·22	22	14	13·17°	14·89°	6·62
11	3	5·5°	4·50°	11·25	23	15	13·3°	14·79°	6·65

6. *Determination of S_g*

$$S_g = f_d S_E + (1 - f_d)\, S_N \qquad \mathrm{W/m^2} \tag{13}$$

where subscripts E and N denote east and north, respectively. From Table III, $f_d = 0{\cdot}391$. The values of S_E and S_N are actual air conditioning loads on the building, arising from direct and diffused sunlight passed through the single glazing, shaded on the inside by Venetian blinds. Building storage effects are taken into account and the values adopted are based on Carrier data.[13] Storage factors from the same source over 24 h are used to extrapolate published air conditioning loads (values of S_E and S_N) for the daytime into the night, when heat absorbed by the floor slab in the vicinity of the window during the day continues to seep into the room after the sun has set.

The term $(1 - f_d)\, S_N$ in equation 13 allows for sky radiation during the time when the sky is clouded over.

The computer results are:

	Unoccupied								Occupied			
Inside θ	0	1	2	3	4	5	6	7	8	9	10	11
S_E	14	11	11	7	7	7	123	177	199	193	158	107
S_N	3	2	2	2	2	1	0	6	16	16	16	16
S_g	7·3	5·5	5·5	3·9	3·9	3·3	48·1	72·9	87·5	85·2	71·5	51·5

	Occupied						Unoccupied					
Inside θ	12	13	14	15	16	17	18	19	20	21	22	23
S_E	73	66	60	54	47	44	38	29	23	20	17	14
S_N	16	19	19	19	19	19	19	7	5	4	4	3
S_g	38·2	37·4	35·1	32·7	30·0	28·8	26·5	15·6	12·0	10·2	9·0	7·3

All values for S are positive and in W/m^2 of glass area.

Since $F^1 = 3421 \times 0{\cdot}5\, (T_w + S_g)$ W, we can evaluate mean energy fluxes in kW during each one hour period (*e.g.* from 0800 to 0900) over 24 h and sum them, keeping gains and losses separate, to give total energy fluxes in the month of April, arising from T_w and S_g, of $-10\,364$ kWh and $+37\,418$ kWh.

7. *Glass flux*

$$T_g = \mathrm{U}^1{}_g (\bar{t}_{eog} - \bar{t}_v) \tag{10}$$

This will be positive or negative according as the sol-air temperature for glass is above (t_{eoga}) or below (t_{eogb}) the mean room temperature ($\bar{t}_r$) for the period considered.

$$\bar{t}_{eog} = \bar{t}_{ao} + R_{sog} f_d E \bar{I}_{Lg} \quad (11)$$

With subscripts a or b added to denote above or below mean room temperature.

$$\bar{I}_{Lg} = -5{\cdot}99\,\mathrm{BK}\left[\frac{273 + 0{\cdot}7\,\bar{t}_{ao} + 0{\cdot}3\,\bar{t}_r}{100}\right]^4 \quad (12)$$

with subscripts a or b added as before. In equation 12, the long-wave emission is a function of glass surface temperature and this has been re-expressed in terms of $\bar{t}_{ao}$ and $\bar{t}_{r1}$ with the assumption that 70% of the thermal resistance of glass is across the inside air film and 30% across the outer.

8. *Determination of* T_g

In April the sun rises at 0500 and sets at 1900 (Table A6.8, IHVE Guide). Processing the meteorological data for April, as outlined earlier, we can establish mean outside temperature values above and below mean room temperatures (Tables I and IV) and their durations, as follows:

Time	I_t	I_{Lg}	$\bar{t}_r$	$\bar{t}_{aoa}$	θ_a	$\bar{t}_{aob}$	θ_b
0500–0800	$\neq 0$	0	19·3°	20·5°	0·12	7·9°	89·88
0800–1700	$\neq 0$	0	21°	21·96°	0·94	10·77°	269·06
1700–1900	$\neq 0$	0	19·3°	20·5°	0·08	7·9°	59·92
1900–0500	0	$\neq 0$	19·3°	20·5°	0·41	7·9°	289·59

The sum of θ_a and θ_b is 720, the total number of hours in the month.

There is some slight error in the approach adopted because $\bar{t}_r$ is not constant at the values shown and neither are $\bar{t}_{aoa}$ and $\bar{t}_{aob}$ constant. Although hourly values of t_r could be obtained by revising the computing programme, with greater complication, the variation of outside air temperature is not known to any better extent.

We can now obtain values for $\bar{t}_{eoga}$ and $\bar{t}_{eogb}$: Since solar gain in the daytime is considered elsewhere (as S_g), the value of $\bar{t}_{eog}$ will only be different from that of $\bar{t}_{ao}$ when I_{Lg} has a value:

Time	I_{Lg}	$\bar{t}_{aoa}$	$\bar{t}_{eoga}$	θ_a	$\bar{t}_{aob}$	$\bar{t}_{eogb}$	θ_b
1900–0500	$\neq 0$	20·5°	19·24°	0·41	7·9°	6·79°	299·79

At all other times $I_{Lg} = 0$ because $I_t \neq 0$.

We can now compute mean values of T_g for the four periods quoted, using equations 10. For example:

1900–0500

$$T_g = 5{\cdot}04\ (19{\cdot}24-19{\cdot}3) = -0{\cdot}302\ \text{W/m}^2 \text{ for } 0{\cdot}41\ \text{h}$$
$$\text{and } T_g = 5{\cdot}04\ (6{\cdot}79-19{\cdot}3) = -63{\cdot}050\ \text{W/m}^2 \text{ for } 299{\cdot}59\ \text{h}$$

0800–1700

$$T_g = 5{\cdot}6\ (21{\cdot}96-21) = +5{\cdot}376\ \text{W/m}^2 \text{ for } 0{\cdot}94\ \text{h}$$
$$\text{and } T_g = 5{\cdot}6\ (10{\cdot}77-21) = -57{\cdot}288\ \text{W/m}^2 \text{ for } 269{\cdot}06\ \text{h}$$

9. *Determination of* F_g

$$F_g = 3421 \times 0{\cdot}5\, T_g \quad \text{W}$$
$$= \frac{3421 \times 0{\cdot}5\, T_g \times (\theta_a \text{ or } \theta_b)}{1000}\ \text{kWh}$$

for each of the four periods of the day considered, to yield monthly totals of + 11 kWh and − 75 034 kWh.

10. *Climatic energy flow through the facade considered*
From the foregoing:

		Negative	Positive
F^1	=	10 364	37 418
F_g	=	75 034	11
F	=	85 398 kWh	37 429 kWh

These values represent 100·72 % and 99·98 %, respectively, of the results printed out by the computer.

Appendix 2—Notation

C	Mean total energy flow rate, in a given period, for the whole building, in a given month.	kW or W
F_1	Mean total energy flow rate, in a given period, through one face of the building, in a given month; subscripts 2, 3, 4 denote other faces.	kW or W
F_R	Mean total energy flow rate, in a given period, through the roof of the building, in a given month.	kW or W
F_I	Mean total energy flow rate, in a given period,	

	arising from natural infiltration through the whole building, in a given month.	kW or W
z	Floor-to-floor height.	m
n	Number of storeys.	–
X_1	Length of any one face of the building; subscripts 2, 3, 4 denote other faces.	m
Q_1	Mean specific energy flow rate, in a given period, through one face of the building, in a given month; subscripts 2, 3, 4 denote other faces.	W/m^2
T_{w1}	Mean specific energy flow rate, in any one hour, through the wall of one face of the building, in a given month; subscripts 2, 3, 4 denote other faces.	W/m^2
A_{g1}	Glazed fraction of one particular external face of the building; subscripts 2, 3, 4 denote other faces.	–
T_{g1}	Mean specific energy flow rate, during the occupied or the unoccupied period, arising from air-to-air transmission and long wave-length radiation exchange with the surroundings, through the window glass in one face of the building, in a given month; subscripts 2, 3, 4 denote other faces.	W/m^2
S_{g1}	Mean specific air conditioning load (heat gain) in any one hour, arising from both direct and diffuse solar radiation through the window glass in one face of the building, in a given month; subscripts 2, 3, 4 denote other faces.	W/m^2
U_1	Thermal transmittance coefficient of the wall of one face of the building; subscripts 2, 3, 4 denote other faces.	$W/m^2\,°C$
$t^1{}_{eo1}$	24-hour mean typical sol-air temperature for the wall of one face of the building, in a given month; subscripts 2, 3, 4 denote other faces.	°C
t_r	Room temperature within the building during a given period.	°C
f_1	Decrement factor for the wall of one face of the building; subscripts 2, 3, 4 denote other faces.	–
t_{eo1}	Typical sol-air temperature for the wall of one face of the building, for each hour of the day in a given month; subscripts 2, 3, 4 denote other faces.	°C
t_{ao}	Mean outside air temperature during any one particular hour in a given month.	°C

R_{so1}	Outside surface thermal resistance for the wall of one particular face of the building; subscripts 2, 3, 4 denote other faces.	°Cm²/W
f_d	Mean fraction of bright sunshine in the daytime or mean fraction of clear sky in the night-time for a given month.	–
α_1	Absorption coefficient for direct and diffuse solar thermal radiation on the wall of one face of the building; subscripts 2, 3, 4 denote other faces.	–
I_{t1}	Intensity of direct plus diffuse solar thermal radiation normally incident on the wall of one face of the building, for each hour of the day in a given month; subscripts 2, 3, 4 denote other faces.	W/m²
E_1	Emissivity of the external surface of the wall for long wave thermal radiation on one face of the building, subscripts 2, 3, 4 denote other faces.	–
I_{LW1}	Long wave radiation emitted from the wall of one face of the building, regarded as a black body; subscripts 2, 3, 4 denote other faces.	–
t_{15}	Mean outside air temperature at 1500 hours, suntime, in a given month; equals the mean daily maximum temperature for the month.	°C
D	Mean diurnal variation of outside air temperature, in a given month.	°C
θ	Suntime from 0 to 24.	h
B_1	Angle factor for the particular building face considered, with respect to the surrounding surfaces; subscripts 2, 3, 4 denote other faces.	–
δ_1	Angle between the building face considered, with respect to the surrounding surfaces; subscripts 2, 3, 4 denote other faces.	Degrees
K	Correction factor arising from the opaqueness of the lower atmosphere to long-wave thermal radiation.	–
p_s	Mean daily vapour pressure of the lower atmosphere in a given month.	mb
U_{g1}	Thermal transmittance coefficient of the glass in one building face; subscripts 2, 3, 4 denote other faces.	W/m² °C
$\bar{t}_{eog1}$	Mean typical sol-air temperature of the glass in one building face during the occupied or the unoccupied	

Symbol	Description	Unit
	period, in a given month; subscripts 2, 3, 4 denote other faces.	°C
U^1_{g1}	Reduced thermal transmittance coefficient of the glass in one building face (omitting the radiant component). Subscripts 2, 3, 4 denote other faces.	W/m² °C
$\bar{t}_r$	Mean room temperature during the occupied or the unoccupied period, in a given month.	°C
$\bar{t}_{aoa}$	Mean value of the outside daily temperature above the mean room temperature, during the occupied and the unoccupied period, in a given month.	°C
$\bar{t}_{aob}$	Mean value of the outside daily temperature below the mean room temperature, during the occupied and the unoccupied period, in a given month.	°C
θ_a	The period of time for which $t_{ao} > \bar{t}_r$, during the occupied or the unoccupied period, in a given month.	h
R_{sog1}	Outside surface thermal resistance for the glass in one particular building face; subscripts 2, 3, 4 denote other faces.	°Cm²/W
E_{g1}	Emissivity of the external surface of the glass for long wave thermal radiation, in one particular building face; subscripts 2, 3, 4 denote other faces.	–
$\bar{I}_{Lg1}$	Mean long wave radiation emitted from the glass during a given period for one particular building face, regarded as a black body; subscripts 2, 3, 4 denote other faces.	W/m²
S_1	Mean specific air conditioning load (heat gain) in any one hour, arising from both direct and diffuse solar radiation incident on the area of sunlit glass in one particular building face in a given month; subscripts 2, 3, 4 denote other faces.	W/m²
S_N	Mean specific air conditioning load (heat gain) in any one hour, arising from solar radiation incident on north-facing glass, in a given month.	W/m²
T_R	Mean specific energy flow rate, during any one hour, through the roof, in a given month.	W/m²
U_R	Thermal transmittance coefficient of the roof.	W/m² °C
t^1_{eoR}	24-hour mean typical sol-air temperature for the roof, in a given month.	°C
f_R	Decrement factor for the roof.	–

t_{eoR}	Typical sol-air temperature for the roof, for each hour of the day in a given month.	°C
R_{soR}	Outside surface thermal resistance for the roof.	°Cm²/W
α_R	Absorption coefficient for direct and diffuse solar thermal radiation on the roof of the building.	–
I_{tR}	Intensity of direct plus diffuse solar thermal radiation normally incident on the roof of the building, for a particular hour of the day in a given month.	W/m²
E_R	Emissivity of the external surface of the roof of the building for long wave radiation.	–
I_{LR}	Long wave radiation emitted from the roof of the building, regarded as a black body.	W/m²
B_R	Angle factor for the roof of the building with respect to its surrounding surfaces.	–
δ_R	Angle between the building roof and the surrounding surfaces.	
$\bar{V}$	Mean specific energy flow rate, arising from natural infiltration, during the occupied period or the unoccupied period, in a given month.	W/m²
m	Natural infiltration rate of air changes.	h⁻¹
h	Floor-to-ceiling height.	m
G	Treated fraction of the gross floor area.	–
θ_b	The period of time for which $t_{ao} < \bar{t}_r$. during the occupied or the unoccupied period, in a given month.	h

Appendix 3—Equations

$$C = F_1 + F_2 + F_3 + F_4 + F_R + F_I \quad (1)$$
$$F_1 = zn\,X_1 Q_1 \quad (2.1)$$
$$F_2 = zn\,X_2 Q_2 \quad (2.2)$$
$$F_3 = zn\,X_3 Q_3 \quad (2.3)$$
$$F_4 = zn\,X_4 Q_4 \quad (2.4)$$
$$F_R = GX_1 X_2 T_R \quad (2.5)$$
$$F_I = n\,GX_1 X_2 V \quad (2.6)$$
$$Q_1 = T_{W1}(1 - A_{g1}) + T_{g1} A_{g1} + S_{g1} A_{g1} \quad (3)$$
$$T_{W1} = U_1 (t^1{}_{eo1} - \bar{t}_r) + f_1 U_1 (t_{eo1} - t^1{}_{eo1}) \quad (4)$$
$$t_{eo1} = t_{ao} + R_{SO1} f_d (\alpha_1 I_{t1} + EI_{LW1}) \quad (5)$$

$$t_{ao} = t_{15} - \frac{D}{2}\left[1 - \sin\left(\frac{\theta\pi - 9\pi}{12}\right)\right] \quad (6)$$

$$I_{LW1} = -5{\cdot}77\, B_1 K\left[\frac{273 + t_{ao}}{100}\right]^4 \quad (7)$$

$= 0$ when $I_t \neq 0$ and *vice-versa*

$$B_1 = 0{\cdot}5\,(1 + \cos\delta_1) \quad (8)$$

$$K = 0{\cdot}56 - 0{\cdot}08\sqrt{p_s} \quad (9)$$

$$T_{g1} = U^1{}_{g1}\,(\bar{t}_{eog1} - \bar{t}_r) \quad (10)$$

or

$$T_{g1} = U_{g1}\,(\bar{t}_{aoa} \text{ or } \bar{t}_{aob} - t_r) \quad (10a)$$

$$\bar{t}_{eog1} = \bar{t}_{aoa} + R_{SOg1} f_d E_{g1} \bar{I}_{Lg1} \quad (11a)$$

or

$$\bar{t}_{eog1} = \bar{t}_{aob} + R_{SOg1} f_d E_{g1} \bar{I}_{Lg1} \quad (11b)$$

$$\bar{I}_{Lg1} = -5{\cdot}77\, B_1 K\left[\frac{273 + 0{\cdot}7\,\bar{t}_{aoa} + 0{\cdot}3\,\bar{t}_r}{100}\right]^4 \quad (12a)$$

or

$$\bar{I}_{Lg1} = -5{\cdot}77\, B_1 K\left[\frac{273 + 0{\cdot}7\,\bar{t}_{aob} + 0{\cdot}3\,\bar{t}_r}{100}\right]^4 \quad (12b)$$

$= 0$ when $I_t \neq 0$ *and vice-versa*

$$S_{g1} = f_d S_1 + (1 - f_d) S_N \quad (13)$$

$$T_R = U_R\,(t^1{}_{eoR} - \bar{t}_r) + f_R U_R\,(t_{eoR} - t^1{}_{eoR}) \quad (14)$$

$$t_{eoR} = t_{ao} + R_{SOR} f_d\,(\alpha_R I_{tR} + E_R I_{LR}) \quad (15)$$

$$I_{LR} = -5{\cdot}77\, B_R K\left[\frac{273 + t_{ao}}{100}\right]^4 \quad (16)$$

$$B_R = 0{\cdot}5\,(1 + \cos\delta_R) \quad (17)$$

$$\bar{V} = 0{\cdot}33\,\mathrm{m\,h}\,(\bar{t}_{aoa} - \bar{t}_r) \quad (18a)$$

or

$$\bar{V} = 0{\cdot}33\,\mathrm{m\,h}\,(\bar{t}_{aob} - \bar{t}_r) \quad (18b)$$

Subscripts 1, 2, 3 etc. denote different orientations.

Appendix 4

THE MODEL BUILDING AND ASSUMED PHYSICAL VALUES

Dimensions

Length: 36 modules × 2·4 m each = 86·4 m.
Width: 2 × 6 m peripheral strips + 1·5 m corridor = 13·5 m.
Height: 12 storeys, each 3·3 m floor-to-floor and 2·6 m floor-to-ceiling.

Glass

0, 25, 50 and 75% on the two long faces but no glass on the two short faces.
Single or double clear glass with internal Venetian blinds drawn to exclude direct sunlight.

Orientation

Major axis pointing N–S, NE–SW, E–W, SW–NE.

U values

Walls: 1·57, 0·96, 0·42 and 0·36 W/m^2 °C
Roof: 1·1 and 0·6 W/m^2 °C
Glass: 5·6 and 2·8 W/m^2 °C for single and double glazing reduced to 5·04 and 2·65 W/m^2 °C, respectively, for calculating sol-air temperatures for windows at night.

Wall types

The four U values quoted correspond respectively to:

(a) 105 mm brick, 50 mm air gap, 105 mm brick, 15 mm plaster.
(b) 105 mm brick, 50 mm air gap, 100 mm light weight concrete block, 15 mm plaster.
(c) 105 mm brick, 50 mm air gap, 100 mm light weight concrete block, 50 mm insulation, 15 mm plaster.
(d) 105 mm brick, 50 mm foamed air gap, 100 mm light weight concrete block, 15 mm plaster.

Occupied period

0800–1700, sun-time, 365 days a year.

Approximate surface densities for solar gains

Walls: 300 kg/m^2 (thickness 250 mm)
Floor slabs: 150 kg/m^2
Roof slab: 300 kg/m^2 (thickness 250 mm)

Decrement factors and time lags

Decrements: 0·3 for walls and roof.
Time lags: 8 h for walls and roof.

Heat transfer coefficients etc.

R_{sow} $= R_{sog} = 0{\cdot}05\,m^2\,°C/W$ normally.
R_{sog} $= 0{\cdot}075\,m^2\,°C/W$ for the reduced U value at night.
α $= 0{\cdot}8$ for walls and roof.
E $= 0{\cdot}9$ for walls and roof.
E_g $= 0{\cdot}925$ for glass.
B $= 1{\cdot}0$ for roof, but $0{\cdot}5$ for walls.
R_{soR} $= 0{\cdot}04\,m^2\,°C/W$.
G $= 1{\cdot}0$.

Appendix 5—Tables

TABLE I
FOR OCCUPIED HOURS OF 0800–1700, SUNTIME

Month	t_r C	m h^{-1}	$\bar{t}_{aoa}$ °C	θ_a h	$\bar{t}_{aob}$ °C	θ_b h	t_{15} °C	D °C
Jan	20°	1·0	—	—	4·02°	279	6·3°	4·1°
Feb	20·33°	0·917	—	—	5·3°	254·7	6·9°	4·7°
Mar	20·67°	0·833	21·58°	1·1	7·9°	277·9	10·1°	6·8°
Apr	21	0·75	21·96	0·94	10·77°	269·06	13·3°	7·8°
May	21·33	0·667	23·3°	17·48	13·67°	261·52	16·7	8·5°
Jun	21·67°	0·583	23·79°	42·71	16·74°	227·29	20·3°	8·7°
Jul	22°	0·5	24·56°	45·56	17·5°	233·44	21·8°	8·3°
Aug	21·67°	0·583	23·87°	43·74	17·28°	235·26	21·4°	8·2°
Sep	21·33°	0·667	23·41°	25·02	15·83°	244·98	18·5°	7·2°
Oct	21°	0·75	23·07°	5·36	12·49°	273·64	14·2°	6·3°
Nov	20·67°	0·833	—	—	7·84°	270	10·1°	4·8°
Dec	20·33	0·917	—	—	4·89°	279	7·3°	3·8°

TABLE II
FOR UNOCCUPIED HOURS OF 1700–0800, SUNTIME

Month	*% Glass*	*Mean inside building temp., $\bar{t}_r$*			
		Heavy-weight		*Medium-weight*	
		Single glazed	*Double glazed*	*Single glazed*	*Double glazed*
January	0	19·7°	19·7°	19·7°	19·7°
February	25	19·6°	19·6°	19·4°	19·4°
March	50	19·2°	19·6°	19·0°	19·4°
	75	19·0°	19·4°	18·6°	19·2°
April	0	20·0°	20·0°	20·0°	20·0°
November	25	19·9°	19·9°	19·7°	19·7°

TABLE II—*contd.*

December	50	19·5°	19·9°	19·3°	19·7°
	75	19·3°	19·7°	18·9°	19·5°
May	0	20·7°	20·7°	20·7°	20·7°
October	25	20·6°	20·6°	20·4°	20·4°
	50	20·2°	20·6°	20·0°	20·4°
	75	20·0°	20·4°	19·6°	20·2°
June	0	21·0°	21·0°	21·0°	21·0°
July	25	20·9°	20·9°	20·7°	20·7°
August	50	20·5°	20·9°	20·3°	20·7°
September	75	20·3°	20·7°	19·9°	20·5°

A heavy-weight building is taken to be 12 storeys high with 200 mm of concrete and screed as intermediate floor slabs. A medium-weight building is similar but with hollow-pot intermediate floor slabs.

TABLE III
BRIGHT SUNSHINE FACTORS AT KEW

Month	January	February	March	April	May	June
f_d	0·187	0·240	0·336	0·391	0·437	0·456
Month	July	August	September	October	November	December
f_d	0·428	0·448	0·402	0·321	0·217	0·186

TABLE IV
DURING UNOCCUPIED HOURS

Month	$\bar{t}_{aoa}$ °C	θ_a h	$\bar{t}_{aob}$ °C	θ_b h
January	—	—	3·0°	465
February	—	—	3·9°	424·5
March	20·8°	0·42	5·5°	464·58
April	20·5°	0·61	7·9°	449·39
May	22·1°	8·68	10·7°	456·32
June	22·8°	29·85	13·8°	420·15
July	23·1°	38·42	14·9°	426·58
August	22·8°	26·85	14·7°	438·15
September	22·6°	8·54	13·1°	441·46
October	21·9°	1·12	10·5°	463·88
November	—	—	6·4°	450
December	—	—	3·9°	465

TABLE V
VAPOUR PRESSURE CORRECTION FACTORS

Month	January	February	March	April	May	June
K	0·343	0·346	0·335	0·326	0·297	0·263
Month	July	August	September	October	November	December
K	0·247	0·242	0·262	0·291	0·312	0·330

REFERENCES

1. Jones, W. P. (1974). 'Designing air conditioned buildings to minimise energy use', *Integrated Environment in Building Design* (Ed. A. F. C. Sherratt), Applied Science Publishers, London, pp. 172–197.
2. Owens, P. (1974). *Energy Budgeting in Buildings*, Pilkington Advisory Service.
3. Jones, W. P. (1975). 'Energy and the Design and Use of Services Installations', R.I.C.S. Summer Conference, Edinburgh.
4. Meteorological Office (1953). *Averages of Bright Sunshine for Great Britain and Northern Ireland*, 1921–50, HMSO.
5. *IHVE Guide*, Book A, 1970, pp. A6–14.
6. *ASHRAE Handbook of Fundamentals*, 1972, pp. 37 and 394.
7. Sutton, O. G. (1953). *Micrometeorology*, McGraw-Hill.
8. Brunt, D. (1932). *Quart. J. Roy. Meteorol. Soc.*, **58,** 389.
9. Meteorological Office (1972). *Tables of Temperature, Relative Humidity, Precipitation and Sunshine for the World*, Part III, Europe and the Azores, HMSO.
10. Bull, L. C. (1971). 'Design for Optimum Building Performance—the Thermal Response of the Building', Haden Young internal seminar, unpublished.
11. Harrison, E. (1956). 'The intermittent heating of buildings', *J.I.H.V.E.*, **24,** 145–187.
12. Jones, W. P. (1973). *Air Conditioning Engineering*, 2nd Edition, Edward Arnold (Publishers) Ltd.
13. Carrier Air Conditioning Company, (1965). *Handbook of Air Conditioning Design*, McGraw-Hill.

Discussion

P. Burberry (*University of Bristol*): In tackling the problem of the form, fenestration, orientation and materials of construction of buildings in relation to energy conservation, Mr. Jones has identified an aspect of crucial importance and one which has received far less than its share of attention because of the difficulty of making accurate prediction. It is to be hoped that Mr. Jones will apply his analysis to more examples and that other people will be encouraged to tackle this very important design problem.

So far the conclusions Mr. Jones has drawn only confirm the design features which designers familiar with the thermal performance of buildings would have expected. The fact that glazing is a dominant feature in the thermal situation, that east/west axes are better for air conditioned buildings and that lightweight leads to energy conservation do not come as a surprise. It would be desirable to have more precise values associated with these phenomena to enable better decisions to be taken during design. It is not really sufficient to analyse a sketch design that has been prepared since there is inevitably a reluctance to change this and start design work afresh. What is needed is preliminary general study of typical problems, using techniques such as Mr. Jones', so that at the very earliest stages of design the broad significance of particular decisions would be appreciated and taken into account.

Two specific points puzzle me in Mr. Jones' paper. Figure 1 shows, if I understand it correctly, a net negative energy flux in June. Since energy flux is defined as the energy flowing into or out of the building through its envelope (including ventilation) a net negative value would appear to indicate that heating is required in June. I imagine that other people would also be puzzled by this point and it would be very useful to have it clarified. The second point arises from Mr. Jones' conclusion that lightweight buildings are more economical from the point of view of energy conservation. This clearly conceals a potential problem of summer overheating. Mr. Jones has concluded that winter energy requirements are greater than summer ones, even in air conditioned buildings, and clearly in non-air conditioned buildings it is only winter requirements which are significant. It is possible, however, to have a building which, while economical in its energy consumption does give rise to periods of overheating in the summer. This is a factor which must be taken into account in design and circumstances can certainly arise where an increase in thermal capacity might be felt desirable to reduce summer overheating or even, in some cases, to reduce the cost or even obviate the necessity for air conditioning plant. The problems of reconciling the winter and summer requirements for thermal performance call for detailed study.

It is worth placing on record that in the University of Bristol mathematical analyses of the thermal performance of buildings have been made using the sol-air method. They have been parallelled by analogue studies which have the advantage of enabling much more accurate predictions of the effect of mass, and the position of insulation to be made and also enable the control, and time-lag characteristics of the thermal installations to be taken into account. It is interesting that these comparisons, while showing a very close correspondence with relatively low thermal capacity buildings diverged more and more as the thermal capacity of the building increased. The analogue value indicating greater economy for high thermal capacity, in the situations studied, than the sol-air predictions. This in no way contradicts Mr. Jones' conclusions about the effect of mass. It may, however, indicate that the excessive use of energy due to mass may be less marked, particularly in summer conditions, than the mathematical analysis indicates.

It is important to conclude by saying that these points are intended as an extension of the ideas put forward by Mr. Jones and are not in any sense a criticism. The problems of the thermal performance of buildings have, except for the notable exception of the development of the admittance method at the Building Research Establishment, been totally neglected for the past 25 years. It is to be hoped that Mr. Jones' notable pioneering contribution will lead to urgently needed and wider studies both by Mr. Jones and many others.

W. P. Jones: The question of the loss in June is answered by our weather. The calculations were over a 24 hour period and the statistical distribution data for London shows that for June the temperature is less than the room temperature during the occupied period (when it is normally steady) and also during the night-time period (when it is decaying) for a significantly sufficient time to give a net energy loss in June.

J. W. Weller (*Adams Green & Partners*): Mr. Jones' computerisation of sol-air temperatures surprises me as sol-air temperatures are an imaginary quantity deliberately fabricated to simplify the tedium of hand calculations. This seems particularly absurd as computer techniques are now available for the calculation of the actual thermal processes occurring in buildings.

I undertook research at Lanchester Polytechnic, sponsored by the British Gas Corporation, on applying such techniques to buildings and this research, which is currently being written up, will enable a thorough analysis of energy control in proposed buildings by the building design team.

W. P. Jones: To the best of my knowledge, manual methods for calculating the unsteady state heat flow through solid walls are only three in number. One method is that developed by Mackey and Wright (and subsequently Stewart) in America, about 30 years ago. At about the same time these people proposed the concept of sol-air temperature and Stewart suggested a simplified approach using equivalent temperature differentials. I am intrigued to know what manual methods you are talking about—although I wonder if they are not analogue studies. The sol-air temperature method is very well established and is particularly useful, as my paper shows, for assessing the climatic impact of such factors as the meteorological statistics for cloud cover upon heat flow through building fabrics. May I add that Peter Burberry's earlier comments support the use of sol-air temperature methods as being substantiated to a satisfactory extent by analogue techniques.

C. S. Weilding (*Surrey County Council*): I was interested to hear Mr. Jones say that double-glazing is not justified for intermittent demands and it follows therefore that this principle applies to the requirements for insulation. There is no point in insulating property to absolute standards when the demand is intermittent which in turn raises the effect of full occupation and part occupation on the economics of energy utilisation.

During my experiences I have found that there are many properties, particularly schools, where owing to deficiencies in such a fundamental matter as zone control, the thermal services for the whole building are in use during evenings and weekends when only a comparatively few rooms are being used. In such cases energy consumption is grossly in excess of the requirement and may be as high as 8:1. Who is to blame for these circumstances? Some blame architects, some blame engineers, some blame planners and others blame those in financial control. It seems to me that the situation must be due to an inherent factor in our systems of planning and procedure because the fact of the matter is that those who are associated in capital expenditure authorisation are also associated—on a posthumous basis—in authorising revenue expenditure. There does not appear to be any legal control on energy revenue and it should not be disregarded that once the energy is used somebody has to foot the bill for it.

Generally, thermal services in schools are designed for full day use and insufficient attention is given to evening and weekend use during the initial planning and subsequent design. Energy utilisation systems for buildings should be designed with a high degree of flexibility, *i.e.* in such a manner that adequate zoning and control make sure during all phases of building use the energy is utilised as economically as practicable.

There is scope for substantial energy savings in existing schools by improving controls and it is thought that the Government may wish to consider making grants for this purpose because it would be a worthwhile investment.

May I comment on optimum start control. One has to remember that a time switch set with an element of shrewdness in conjunction with automatic control can do a very good job. If the energy input is a bit lean during initial occupation of the building the timer will save energy because the optimum start control in such instances would bring the plant on earlier and therefore use more energy. This factor must be taken into account when assessing the economics of installing an optimum start control because it may be more costly to run than other well established methods already being used.

On the question of heat pumps, one must not forget the Festival Hall experience which was disastrous—it was overcalculated and to set a gas fired boiler efficiency at 60 % when comparing it with a heat pump is unreasonable (Table 6, p. 174). Good gas fired boilers should yield 75–80 % efficiency which indicates an error approaching 50 % in the savings advocated for the heat pump. If such misleading figures are published without question there are bound to be some erroneous calculations with subsequent dire results on the economy.

I also find much to my concern that swimming pools, also temporary, permanent and speculators buildings use electricity extensively for heating and in many instances not even off-peak tariffs are employed. Those who understand energy values must find such an occurrence to be a very serious matter and a gross misuse of a high class, high price energy. I think that the Government should place some embargo on this practice which is to be deplored.

Dr. M. A. Osei-Bousu (*Newcastle upon Tyne Polytechnic*)*:* Is Mr. Jones advocating a new method for calculating heat loads, and how accurate is his method compared with the other methods being used? Can Mr. Jones give us an indication of the percentage accuracy of his method?

W. P. Jones: I am not advocating new methods. I have been using entirely traditional methods but with typical weather data instead of design weather data. I do not think it is necessarily more accurate. The accuracy depends upon the accuracy of the assumptions made and on the accuracy of the weather data to hand in a particular case.

G. S. Vincent (*Building Design Partnership*)*:* Has Mr. Jones extended his method to achieve a calculation of the energy consumption throughout the year for the building and as part of the development. He will be able to take into account the variable indoor air temperatures, particularly in the summer, so that one can, from this method, assess this effect on the annual energy consumption.

W. P. Jones: Yes indeed, the method does calculate the energy consumption over the year and it can be applied to any building. It is available for use right now; in fact, with the appropriate modifications it can take account of variations in room temperature too.

A. Newton (*Pilkington Bros. Ltd.*)*:* I am interested in Mr. Jones' comments that positive and negative heat flows can never be regarded as cancelling out, implying that all heat gains must be treated as an unmitigated nuisance. Whilst such an approach could be justified in an air-conditioned building for control reasons it is not applicable to buildings in general where winter sunshine through glazing can provide useful heating, see Davies* and the comments of Nicklin (pp. 209–17). This is not to suggest that there is a proven case for the unrestricted use of glazing to save energy but simply to point to evidence that the window can make a positive contribution to the energy balance of a building. There is good reason therefore to refrain from adopting the 'rule of thumb' from Mr. Jones' general conclusions which is, 'the less glass the better' for the thermal performance of the building.

On the subject of 'rules of thumb', these might have been necessary before modern information processing techniques became available but now there is a chance to do something better. Computer programmes are known well enough in their role of performing detailed calculations given the time and trouble to prepare and analyse the data; however this is of limited use to a designer who needs a quick evaluation of possible alternatives. Attention must be paid, therefore, to another way of using the computer which could eliminate the need for rules of thumb. A number of interactive programmes have been developed for building design which perform the necessary calculations quickly and with the minimum of data preparation and communicate with the user by question and answer at a typewriter-style keyboard. The availability of portable terminals permits the architect or engineer to use the computer from his office as a working design tool without the need to purchase or install any specialist equipment.

W. P. Jones: I first of all feel that you may have misunderstood what I have written if you imagine that it is based on a rule of thumb. Somewhat to the contrary, I took pains to avoid using 'thumbs' and in fact I used a good deal of data that was provided by Pilkingtons in assessing the penetration of daylight. The first point, I think also you have misunderstood. If you look at the text you will not find that I described sunlight, or heat gains as 'an unmitigated nuisance' to quote your own words. My point was that heat gains do occur and heat losses do occur and using the best data available one cannot say that they will occur simultaneously. Now we know of course that they sometimes do occur simultaneously and as you rightly observe heat gains are very often a benefit, but we cannot jump to the conclusion that heat gains will always cancel heat losses. Finally, also if you refer to my paper, you should see that I did not jump to the conclusion that minimum glass was the absolute answer. I

* Davies, M. G. (1975). Heating buildings by winter sunshine, *Building Science.*

pointed out that there is some evidence that about 25% glazing is a beneficial answer in thermal terms and if we accept the human element as well, I suggested something like 25–30% would be a good proportion.

A. Newton (*wrote*): Whilst not calling gains an unmitigated nuisance Mr. Jones has treated them as such by proposing the gross energy flow as an indication of building energy requirements. It is true that he considers the benefits of windows in providing light and so saving energy on electric lighting (hence his indication that 25% glazing might be desirable). However, the window is still treated as a loser of heat because gains and losses are added arithmetically, and so the potential of the window to make a positive contribution to the building's heat balance is largely ignored.

If such a simplification is used as the basis for a sophisticated analysis then the conclusion of that analysis, expressed in the succinct form that brevity requires, can have only a limited validity. This is the basis of my comment on 'rules of thumb', not to suggest that Mr. Jones has used one but that he has possibly produced one, creating an impression of certainty in a field that is still open to argument.

W. P. Jones (*wrote*): I can understand Mr. Newton's very obvious concern about the use of glass and I sympathise with him. Nevertheless the analysis points clearly to the conclusion that the less glass used the better, no matter whether one takes losses or gains in isolation or whether one cancels them. Cancellation helps greatly when it can be taken advantage of. Unfortunately, as I was at some pains to point out, the prediction of cancellation is uncertain and it was for this reason that I made the tentative proposal of grossing up energy fluxes. It is also true, as Mr. Billington later points out, that advantage can be taken of electric lights (with some greater degree of assurance than glass offers) to cancel heat loss. Also very much to the point is the comment of Mr. Dubin below that advantage can be taken of solar gain through glass during daylight on Southern exposures, without the excessive penalty of night-time loss, if external shutters can be closed at night to improve the poor U value of glass.

The conclusions quoted in the paper are very obviously only strictly applicable to the limited case of the hypothetical building used as an example, although evidence from other sources supports the general contention in part, as Peter Burberry remarked earlier. The analysis and programme put forward in this paper is a flexible one and will permit examination of the many different building options possible.

N. S. Billington (*Building Services Research and Information Association*): I found Mr. Jones answer to Mr. Newton somewhat disappointing. I understand his logic of course, in adding the positive and negative contributions in an arithmetical way, but I have some doubt as to the legitimacy of this when we are talking about internal heat gains, such as lighting. In winter, one knows the light is going to be on, and one knows that for most of the time the outside temperature is going to be lower than that which you need inside. The lighting energy will certainly make some contribution to the heat requirements, and so I am very sceptical of the conclusion that Mr. Jones comes to that low U values are always beneficial, even with high lighting levels.

W. P. Jones: What Mr. Billington said is true, but there was a difficulty about what one should do with the answers obtained. It is pretty easy to use a computer and get drowned with reams of paper and masses of figures, but what do you do with it at the end? My tentative proposal was to add the positive and negative to offer a gross answer. I agree that there are refinements possible in the treatment.

May I add that the method I outline gives the answers concerning factual heat flow into and out of the fabric, before any system is proposed. The interpretation of these figures, as with any building energy index, is a matter of refinement. In any case, cancelling the lights completely would be unfair, since they are off at night and my method covers 24 hours. I agree some cancellation might be considered for the occupied period in establishing a Lit Building Energy Index.

F. S. Dubin (Dubin, Mindell, Bloome Associates, P.C.): In the UK climate single glazing on the southern facade of the building arranged so that it can receive the winter sun when it is available, with thermal barriers to close it off at night would, I submit, be an energy conserving design. I would disagree that 20–25 % glazing arbitrarily is something to consider—you must consider the orientation of the facade. On double glazing in this climate, on the northern exposures particularly, and the east and the west next, adding a double sash to an existing single glazed surface would decrease infiltration, which may be even a greater benefit, then the reduction in conduction losses, adding that layer outside, with reflective glass may provide additional cost benefits all the year round and should be considered in an analysis.

A good point on alternative sources of energy that has been made recently in the press is that the effort to perfect the production methods for developing photovoltaic solar cells could have a tremendous 'payback', and by the year 2000 cells could produce anywhere from 30–100 % of all electrical energy requirements. I believe the figures are well founded.

W. P. Jones: The method I have proposed does deal with the building with different orientations and the results I gave were for fairly simple glass assumptions, and the percentages referred to the building as a whole. I suspect that if one looked at the southern exposure in more detail with the assumption of external shutters at night, one would get rather interesting answers for that aspect and show that what you say is true.

D. M. Andrew (Vice President National Council of Maintenance Associations—National Westminster Bank Engineers Group): I was astounded to hear from Mr. Weilding that some swimming baths are electrically heated. This is a clear case for a useful suggestion to the Department of Energy that this situation should be looked into, because there are hundreds of swimming baths! There is a good case here for using gas-fired boilers, where the waste heat is clean. A lot of work has been done in France on this subject,* total heat efficiencies in the order of 95 % are claimed. (The low grade of waste heat available in this instance matches quite well with the grade of heat needed to reheat the recirculated and filtered swimming bath water.)

H. C. Jamieson (*Haden Carrier Ltd.*): Mr. Jones has not emphasised the importance of his paper to the extent I think he should have done. It really is quite a different approach to the design of buildings.

Any designer of mechanical air conditioning installations for buildings, knows what it is like to be given a plan of a building which is grossly wasteful in energy. Peter Jones' approach is to assess the building envelope on its own as to its energy requirements before it has got to the stage of going to mechanical and air conditioning engineers. This is the importance of his approach.

* Building Services Engineer, Volume 42, September 1974, pp. A26/A27.

13

Building Services—An Energy Demand Review

J. C. KNIGHT and L. G. HADLEY

Synopsis

The ways in which energy is used in the two principal building sectors, namely domestic and commercial/industrial, are reviewed and the future trends of energy demand analysed. The use of design and annual consumption targets are referred to, in so far as they are likely to affect energy demands for new buildings. How energy policies may also affect demands is discussed.

Introduction

That the demand for energy, by engineering services in building, is a significant part of the total national requirement, has been demonstrated and emphasised in most of the discussions on energy conservation which have taken place over the past few years. In many of the papers presented to this conference, the point has received further emphasis.

If the demand of all the different types of buildings is expressed as relating to units of floor area (a convenient if not particularly accurate guide) as, for example watts per square metre (W/m^2), then the quantity varies widely according to the shape, and use of the building and the complexity and sophistication of the engineering systems serving it.

A comparison of net and gross energy consumption by the domestic and industrial/commercial sectors is shown by Figs. 1 and 2. While part of the gas, oil and electricity in Fig. 2 is no doubt used for industrial processes and manufacturing, it is clear that a substantial proportion is represented by energy used in buildings.

The way in which the net and primary energy is consumed in the

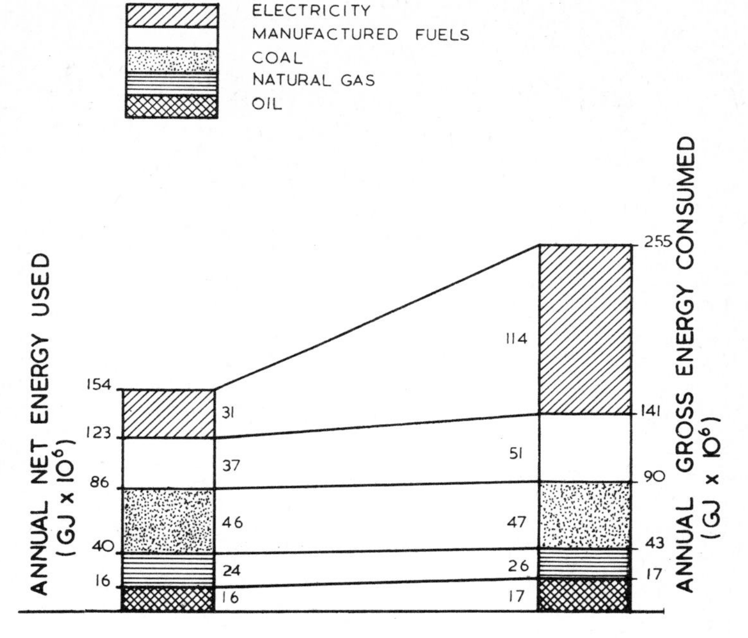

Fig. 1. Nett and gross energy consumption for domestic sector.

various categories of commercial and Public Service buildings, (excluding factories) is shown in Fig. 3. However, the energy demands of the future generation of buildings are likely to differ significantly, due to our present consciousness of the importance of energy conservation.

The Domestic Sector

While the current total energy demand of the Domestic Sector is probably known with a greater degree of accuracy than any other, forecasting future trends in demand of the various energy sources is fraught with greater uncertainty. The exploitation of the North Sea gas fields has brought about revolutionary changes in the ways in which houses are heated. Equally important changes are likely to occur, due to realistic pricing policies relating to the use of electricity and light fuel oils.

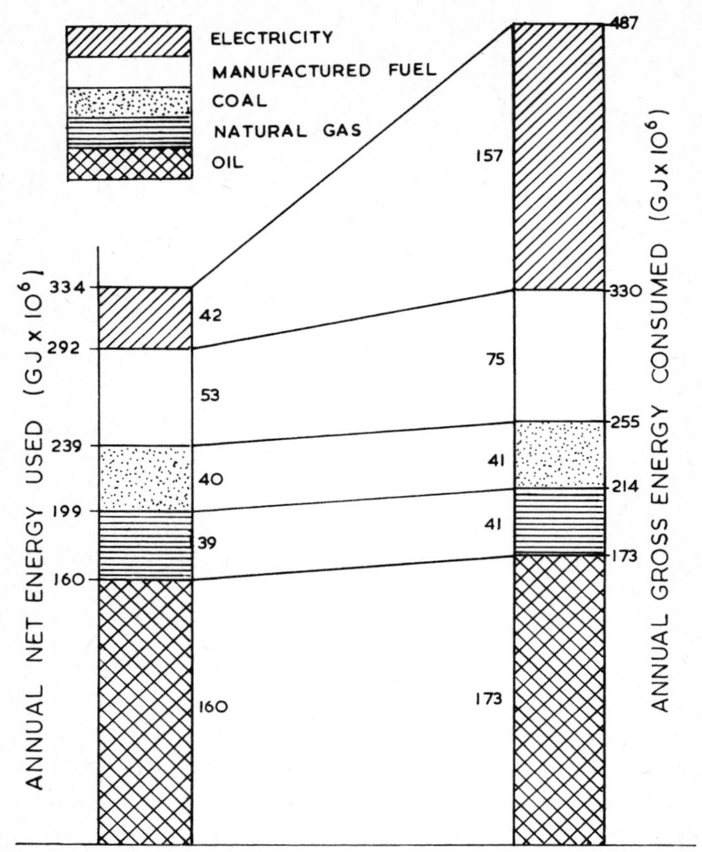

Fig. 2. Net and gross energy consumption for industrial and commercial sectors.

At the present time the high proportion of the domestic energy demand, represented by electricity, is a consequence of a commercial policy to increase 'load', and a political policy in the interest of maintaining cost of living indices. It has also been the only simple way of providing individual room heating for the greater part of our domestic buildings which were originally designed to be heated by open coal fires. The uneconomic use of raw fuel, resulting from these policies, is in the process of being rectified to some extent by the new tariffs.

Reserves of natural gas are finite, and on a world scale significantly less than those of oil. A policy of conservation of natural gas, perhaps

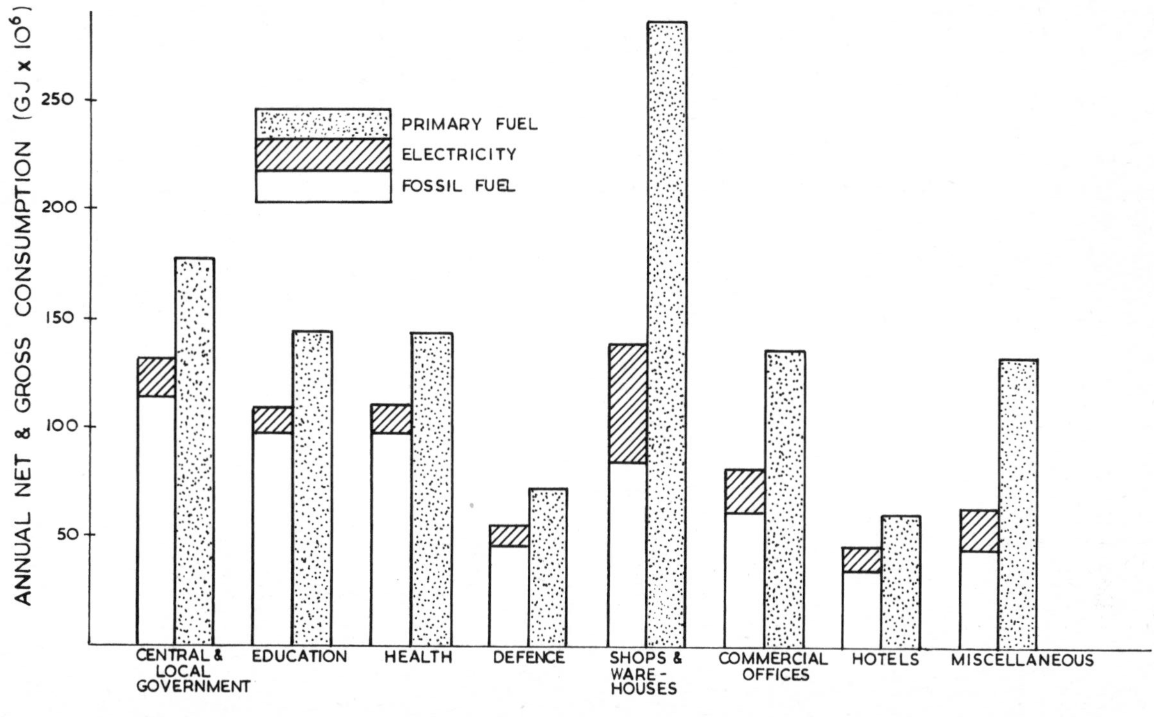

Fig. 3. Estimated net and gross energy consumption in the public service and miscellaneous categories.

of attempting to limit annual consumption rates roughly to the present level, will have a direct influence on trends in domestic consumption. Experience in the USA of the growth in demand for natural gas, by industry as well as the domestic consumer, and the consequent rapid decrease in reserves are a reminder of what can happen in the UK. A 45 % increase in gas demand in the USA over the past 10 years, has coincided with a run down in production, due to known reserves being lower in spite of greatly increased exploration and exploitation of indigenous resources.

The chances of coal reasserting its original position, as a main fuel source for the domestic sector, would seem to be outside the realms of practical possibilities if only because of the atmospheric pollution problems associated with its use. Manufactured solid fuel from coal does not offer a solution to the problem of greatly increasing domestic coal consumption because of cost of production, relatively high overhead losses, and the environmental and atmospheric pollution problems created by the large processing plants required.

The way ahead to making greater use of coal, lies in the conversion of existing coal fired electricity generating stations to the production of both heat and power together with the development of new district heating principles; the building of new combined heat and power stations, fired by coal and located near to the city and the load centres of gravity; direct coal fired district heating plants for new towns and city centre developments.

It can be argued, that the UK coal production capacity must be secured and increased, not only in Britain's interest but that of the European community as a whole. Since the British coal industry produces, at the present time, mainly steam-raising coals, positive progress in adapting existing generating stations to combined heat and power production, plus concentration of district heating development in the domestic sector to coal-fired plants, will have most important long term benefits to the overall economy of the EEC.

But whatever the energy sources may be, what of the size of the future demand? The housing stock in 1973 was approximately 20 million, of which 52 % were owner occupied, 17 % rented from private owners and 31 % rented from local authorities. The number of new dwellings erected in the same year, was about 300 000 a number not greatly varying over the past 2 decades.

The growth in the housing stock is therefore around $1\frac{1}{2}$% per annum, which presumably hardly keeps pace with the decay, and can be ignored for all practical purposes of predicting future energy demands. It is questionable therefore, whether improving the thermal insulations of dwellings will of itself effect a corresponding reduction in the total energy demand of the domestic sector. So many of our existing houses are at present inadequately heated, and equipped with systems of insufficient capacity for whole-house heating, that improved thermal insulation is likely mainly to produce improved environmental standards. Consumption can be reduced, but not to the extent indicated by the change in the rate of heat requirement.

This is confirmed by an examination of published statistics for 1973, in which housing accounts for 64% of total building stock, expressed as 1324 million m^2 floor area. If the final consumption of all forms of energy in the domestic sector is compared with this figure, it appears that energy is consumed at the rate of about 11·0 MJ/m^2. Similarly, it can be deduced that the average dwelling is of about 66 m^2 floor area, and consumes approximately 730 MJ/per annum.

Since the amount of heat required to achieve Parker Morris standards (and as recorded for Local Authority district heating schemes) is about 650 MJ, it does appear that with the present consumption much better indoor conditions could be achieved by a combination of improved insulation, weather stripping and more responsive and effective heating appliances which, in the end offer annual energy savings of 100–150 MJ per dwelling.

This is not an argument for going slow on improving thermal properties of houses. On the contrary, the demand for improved environmental and health conditions will, unless it goes in step with improved insulation etc., lead to greater demands for the refined fuels—gas and electricity—than the country can afford.

In fact, given current standards of house heating in the UK, an average consumption of around 730 MJ per annum indicates a very extravagant use of energy. Admittedly, the district heating figures do not include lighting and cooking, but taken as an average for the 20 million dwellings concerned, this should not exceed about 100 MJ per annum for each dwelling. The figure of around 650 MJ per annum occurs in most studies of domestic heating in the UK.

For example, the South London Consortium was recently reported to be making a plea for local authorities to become conscious of the need to conserve energy, and asked for standards of insulation to be improved beyond those recently revised in the Building Regulations. In support, the Consortium quoted that the 127 700 dwellings of the combined local authorities consumed 508 000 tonnes of coal equivalent per annum. At about 4 tonnes per dwelling this is equivalent to 650 MJ (approx.) per annum, and the figures were used to support a call for experiments with 'renewable sources of energy such as solar and wind power'. How much more worthwhile, however, would be experiments to reduce consumption in the existing dwellings.

It is now practicable to design new local authority houses which should achieve whole house heating, with a consumption of only about 300 MJ per annum. The real problem facing local authorities and the country generally, is how to modify existing dwellings and their heating systems, to approach the modest fuel consumption now capable of being achieved by modern house designers.

So far as the new stock is concerned, it is occasionally argued, that insulation to the Standard of the new Building Regulations (and particularly so when even better standards are worked to) will result in rather lower thermal efficiency of heating appliances, because of their lack of rapid response to changes in rates of heat requirements, and that quick response electric heaters with virtually 100 % efficiency are therefore justified on a primary fuel usage comparison. This does not take account however of the enormous additional capital expenditure on power stations, and the distribution network eventually committed to by increased on-peak electricity load and the argument is invalid on that account if no other. The way to better use of energy resources in the domestic sector is to remove the electric space heating load, to work for better thermal efficiency and quicker response of fossil fired appliances to changing loads.

It seems reasonable to conclude that in spite of the possible theoretical savings, final energy demand by the housing sector should remain at the current rate, but substantial reduction in the total of primary input energy will only be achieved by major changes in fuel policy. For example, by switching from electricity to fossil fuels and/or development of large scale coal fired district heating and combined heat and power generation.

The Commercial and Industrial Sectors

The non-domestic sector of our building stock represents about 35 % of the total floor area, but it is likely to account for a greater proportion of that part of the total energy consumed in buildings, due to different heating standards, greater use of electricity for lighting and power, larger space per unit floor area and similar factors.

Office buildings account for about 70 million m^2 of floor area, 3·4 % of the total, while industrial buildings including shops and warehouses amount to about 20 % of the total building floor area or about 400 million m^2.

Earlier in Fig. 3 we illustrated how the energy consumed by these different categories of buildings is shared. The effects of converting the energy into units of primary fuel input is also shown, underlining the penalty in these terms of using electricity or manufactured fuels.

In the Government statistics, shops and warehouses are lumped together, but the over-riding importance of the artificial lighting load can be seen from the diagram and it is to this area of energy demand that attention has to be given if any meaningful reduction is to be achieved.

Taking the breakdown of building categories suggested by the current statistics, and assuming arbitrarily the rates of ventilation normally considered necessary for the various types of occupation and usage, it is possible to produce a picture of the main factors affecting energy use. This is shown in Fig. 4, and indicates how important a contribution to energy demand is represented by the ventilation rate. In the buildings used for Education and Health purposes, it can be observed that only by controlling ventilation to lower standards can consumption be materially reduced. Tampering with existing thermal insulation standards will not have any noticeable effect.

The demands for energy by engineering services other than heating and lighting, are shown typically by Fig. 5 and are relatively insignificant.

Clearly, industrial building other than shops and warehouses, comprise about 255 million m^2 of floor area or 12·5 % of the total building stock of the country. This represents a block of buildings worth special attention, not only because it is a substantial user of energy, but also because this type of building has high rates of heat loss per unit of floor area. The Thermal Insulation (Industrial Buildings) Act has been a contributory factor in this, requiring as it does reasonable insulation standards to be applied to the roofs of industrial buildings. Factory type buildings frequently have rates of heat requirements of around 200 W/m^2 area as compared with suggested 100 W/m^2 for similarly shaped offices.

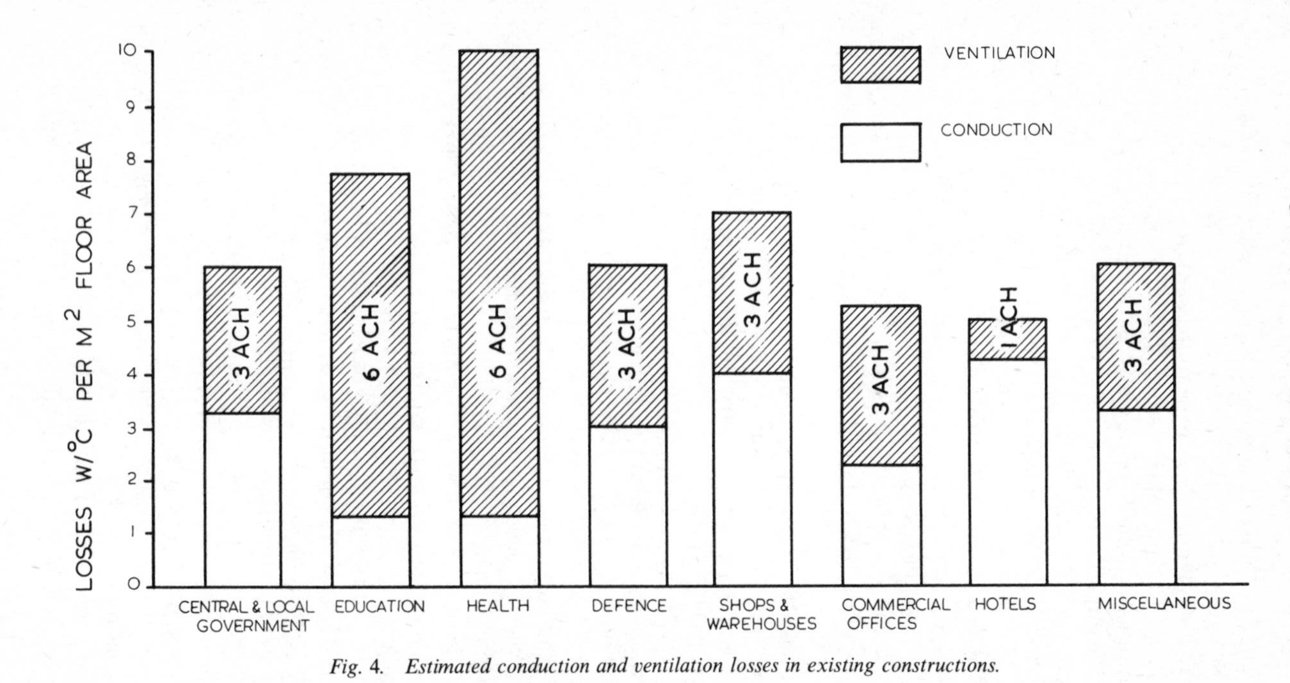

Fig. 4. *Estimated conduction and ventilation losses in existing constructions.*

Improving the thermal insulation of factory walls to comply with the new regulations for domestic buildings, could possibly reduce present rates and thus theoretically produce a reduction in energy used.

But the problem is not quite as simple as it might seem. Improving thermal insulation of the walls will be a costly and fairly difficult

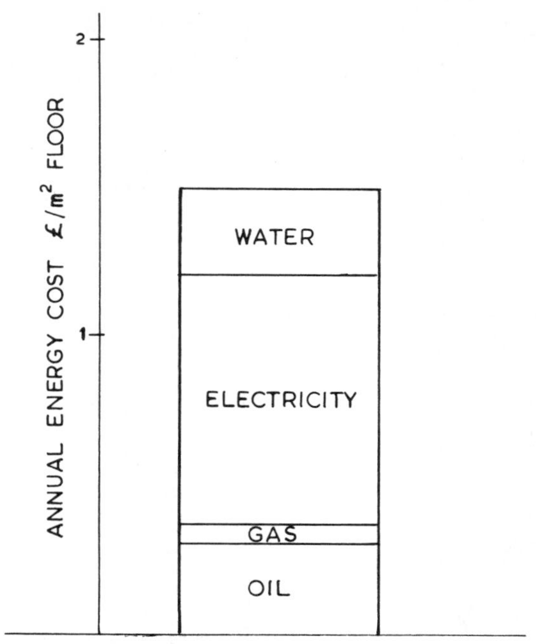

Fig. 5. Typical annual engineering cost for heated office buildings. After BRS Digest 138.

procedure in existing buildings, and the cost will vary widely according to the original design features. Also the smaller fabric to space ratio, typical of factory buildings, means that differences in U values of the fabric have little effect on the overall heat demand.

A factor worth considering is whether reductions in fuel demand can be achieved by adapting or renewing the type of heating system to make it more suitable and more responsive to need. This, in conjunction with modern control systems, might be a more profitable approach to fuel economy than spending a great deal of money improving the thermal insulation of the walls—especially when the comparatively short period of occupation of factory buildings, with their low thermal capacity, is taken into account.

Office buildings are probably the most interesting of all the categories of buildings, because of the sophistication of the engineering services now required, and the direct impact services have on the value of the property, and on the efficiency and comfort of the people who work in the buildings. In the total terms of energy usage, however, this class of building is by no means as important as it first appears. Office buildings account for only 3·5% of our total building stock, and about 70 million m^2 of floor area. It can be argued, however, that the modern office building with its complicated air conditioning and other sophisticated engineering services is a prodigious user of energy. Certainly, in no other type of building does energy consumption vary so widely, and in no other case does the choice of heating and air conditioning systems so affect the efficient use of energy. The energy consumption of a centrally heated office building is frequently only 30% of that of a similar building which is air conditioned and illuminated at 'prestige' levels. Even for similar types of air conditioned buildings, the choice of systems and care in design details can produce a difference in energy consumption of 2:1, and even higher ratios of capital costs and of primary energy used.

However, the rate of provision of new air conditioned office accommodation, is probably only about 1 million m^2 floor area per annum as against, say, a total of 30 million m^2 floor area of all types. Also, the amount of air conditioned office area now in use is probably only 5 to 10 million m^2 compared with 60 million m^2 of centrally heated accommodation. The importance of the new centrally heated stock of buildings is evident if any determined onslaught on existing rates of energy demand is to be effective. Fortunately, from this point of view, the choice of a heating system and its design features have little overall effect on energy consumption of centrally heated offices. What is much more important (assuming fortuitous ventilation due to faulty windows and doorways has been dealt with) is the way in which the heating system is operated—its firing system, choice of fuel and its ability to respond to the regime of occupation. This was demonstrated many years ago by Knight and Cornell (1959),[1] and Knight and Jones (1963)[2] and with the studies of Harrison (JHIVE 1956),[3] and Colthorpe (JHIVE 1964)[4] eventually led to the development of the Optimum Start Control technique, developed by Jackson (JIHVE

1971)[5] and introduced on an experimental basis to the Property Services Agency of DOE circa 1969. The subsequent application and development is touched upon by Johnston and Gronhaug earlier in this conference.

Because of the over-riding importance of choice of fuel, mode of operation and the relative unimportance of the type of heating system, it has been possible to develop some indices of fuel consumption, for example, kW/w of external fabric heat losses or, MJ/m^2 of floor area in order to predict or compare the energy consumption of buildings.

Before any real progress can be made in reducing energy demand in particular cases, we must have some agreed norms for fuel consumption of the different classes of buildings, which take into account the characteristics of the building envelope and the part of the country concerned. Energy audits can only be made if this kind of information is available. Attempts to establish 'norms' of fuel consumption for the various types of constructions, installations and modes of operation of heating systems have been made over the last 15 years of so but none, until recently, has fallen into common usage. The 1959 work of Knight and Cornell,[1] for example produced a target figure of fuel consumption for centrally heated office buildings, intermittently operated, viz.—3·40/kg oil/degree day/therm of heat loss. An alternative approach developed further by Knight and Jones[2] employs the use of E values originated by HVRA, representing hours of operation at the design heat loss kWh/kW.

Values of E were derived for most of the regions of the UK and they varied in the case of intermittently oil fired heating systems from 1300 in Norfolk to 1800 in Scotland. These E values allow fuel demand to be predicted for any of the areas, by merely taking the appropriate seasonal mean value of E and by multiplying it by the design heat loss of the building to obtain the annual fuel consumption. The importance of designing for a regime of occupation is shown by the fact that E values established for a continuous mode of operation, range from 2100 to 3100 as compared with intermittently operated systems varying from 1300 to 1800. The value of this approach to predicting or comparing fuel consumption is that the minimum fuel consumption can be determined quickly and comparisons readily made with various modes of operation and energy sources. The use of E values also allows fuel consumption to be measured against the optimum theoretical figures.

However, these methods of predicting fuel consumption while providing a national basis for comparing existing buildings do not of themselves provide design target figures for new buildings. The use of E values and the units of fuel consumed per degree/day do not in the end determine fuel consumption per unit floor area, since they depend upon the calculated heat losses from the building—however great they may be. To ensure a constant economy is observed throughout the overall design stages of the building and its services we need energy targets (W/m^2 floor area) for various building shapes and types.

An attempt to establish such energy targets has been made by Cornell and Scanlon[6] of PSA and the results of their work for naturally ventilated and centrally heated offices will be published, but we have their permission to illustrate how such targets can be used by designers and/or clients. The targets indicate the initial energy standards against which design comparisons of buildings and services can be made for any particular building requirements.

Firstly, a Design Energy Target is set. This limits the amount of energy to be used in W/m^2 floor area, the figure varying according to building shape, size, location etc. and expressed as a steady state maximum (*i.e.* design heat loss and allowance for lighting etc.)

Secondly, Annual Energy Targets are set. These are the anticipated gross energy consumption figures in MJ/m^2 based upon the energy targets and calculated to take into account climate, fortuitous heat gains, intermittency of operation and system efficiency. The targets can be applied to future buildings and are based on construction to the new domestic thermal insulation standards of the 1975 Building Regulations. They can also be used to monitor consumption in existing buildings, in which case they are based upon preceding thermal insulation standards with windows occupying between 30% and 70% of the external vertical fabric.

The targets can also be used to compare the energy performance of different building designs and shapes having the same floor area. For example it can be assumed that the optimum shape of an enclosure for a minimum heat loss is a square plan with a height of half the side. But the more usual shape resulting from other factors such as requirements of natural daylight and ventilation will normally have a width of 15 m and using this as the main criterion, Cornell and Scanlon[6] have developed maximum energy targets for a range of floor areas and their corresponding optimum shapes.

By developing the equation:

$$Ao = \frac{81\, U_{v1} n^2}{U_H}$$

where: Ao = Occupied floor area.
n = Number of storeys.
U_{v1} = Thermal transmittance of the vertical surface whose width is 15 m.
U_H = Thermal transmittance of the horizontal surfaces.
they provide for a ready assessment of, for example the maximum number of floors and maximum length of a side for any shape. The formula may also be used to compile the optimum shape and hence the energy target for any combination of thermal properties. Examples of the use of the information are as follows:

Given a particular building shape, the ratio of floor area to area of exposed fabric is calculated. From the tables of Energy Targets against the floor/fabric ratios the appropriate targets are obtained.

Another set of tables gives 'Maximum Energy Targets from Minimum Thermal Envelope', from these the optimum ratio can be read against the energy target figures to compare thermal efficiency.

Widespread use of energy targets depends upon their versatility. That is to say they must be capable of being applied to any shape of building and whether it is free standing, has one two or three walls exposed and irrespective of height. It is also important that their application to new designs result in some kind of discipline such as ensuring that both envelope and engineering design are energy conserving. Given that the wall transmittance is 1·0 W/m^2K and the average value for walls plus windows is constant at 1·8 W/m^2K it becomes practicable to produce various ratios of building fabric area to floor area—Wall/Floor or the alternative, Floor to Wall. In the event, the latter is chosen and targets set out relating to increasing areas of floor to unit fabric area.

Given a particular shape and number of exposed walls (free standing or terrace) then as the number of storeys increases so the ratio increases. Thus a small building, single storey 15 m × 15 m will have a ratio of 0·85 if free standing. If with the same ground floor area the number of storeys is increased to 6 the ratio becomes 1·24 (approx.). Generally, the low ratios are for small free standing rectangular buildings and, these get higher as the building height increases. High ratios apply to buildings with limited area of exposed wall, *e.g.* built adjoining others.

In calculating the figures, and computing the fabric area, allowance has to be made for losses through the floor and roof and a rate of air change of one per hour is assumed. Judgement has also to be exercised regarding illumination standards, periods of occupation and seasonal

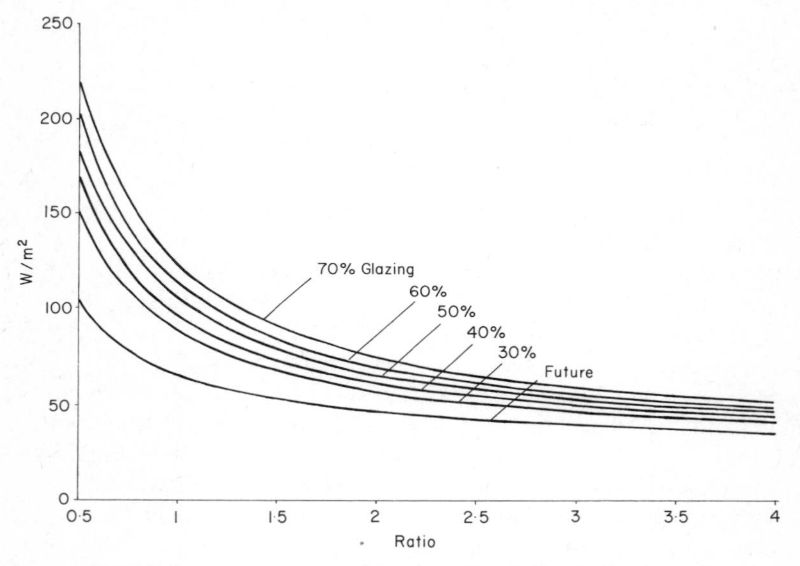

Fig. 6. Design energy targets for naturally ventilated office blocks.

efficiency of the boiler plant. Given these basic assumptions a family of curves can be developed as shown on Figs. 6 and 7 for both existing and future buildings. Figure 6 gives Design Energy Targets for office buildings and Fig. 7 gives Annual Energy Targets. Similar curves can be produced for other building types with different periods of use, rates of natural ventilation and applying other assumptions regarding lighting standards and thermal efficiencies of engineering plant.

That such targets can be made to work, and are reliable indicators of energy demand is borne out by comparing the Cornell and Scanlon[6] targets for office buildings with the earlier Knight and Cornell[1] figures of fuel consumption obtained from analysing records of consumption of some fifty Government office buildings with other similar published figures. The fuel constants so developed and

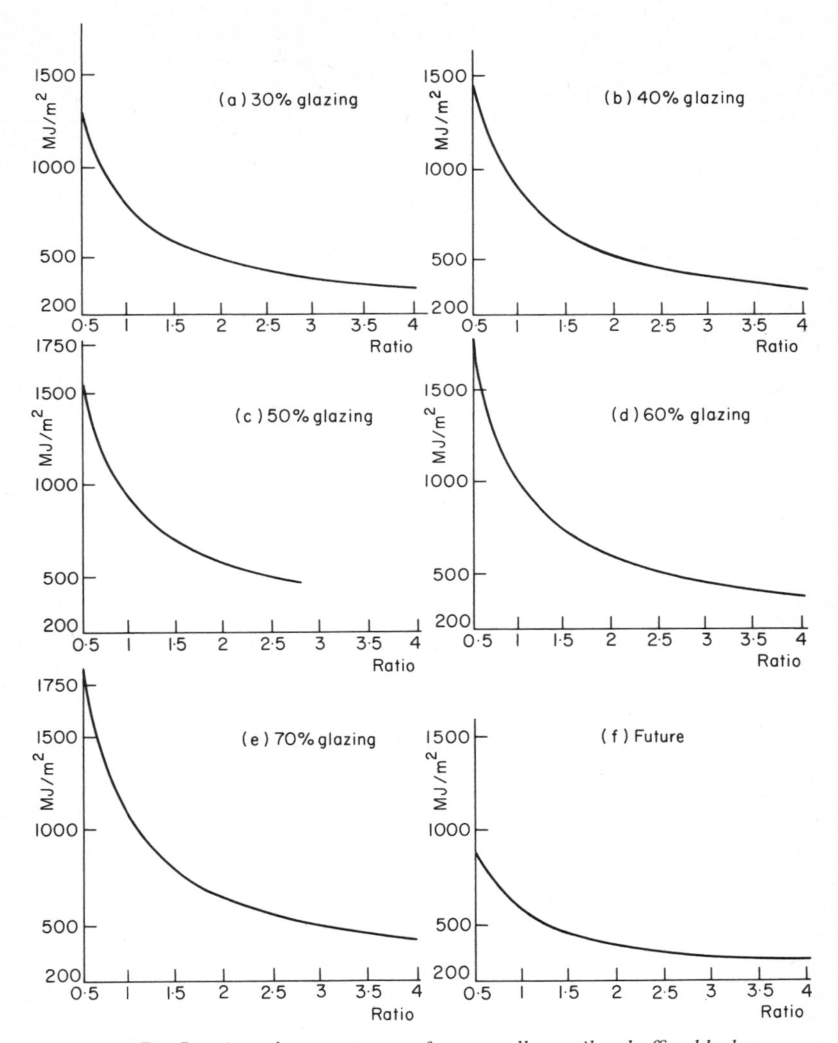

Fig. 7. Annual energy targets for naturally ventilated office blocks.

expressed in, for example, kg of fuel/degree day/therm of heat requirements can be directly compared by taking the constant for intermittently operated oil fired systems. This when converted produces a figure of 6·05 MJ consumed per year for each watt of heat required. From the Cornell and Scanlon[6] curves where the building ratio is 0·5 the target design rating for ventilation plus fabric losses is only 96 W/m^2 and the annual consumption given as 794 MJ/m^2 of heat requirement. Where floor area to external weighted fabric losses is 4·0 it can similarly be deduced that energy consumption should be 6·13 MJ/W of heat requirement. This latter figure compares closely with that of 6·05 MJ/W referred to earlier. No more direct comparison is possible because the idea of comparing buildings on a floor/wall ratio had not been considered during the early investigations and average heat loss per unit of fabric area was considerably higher than now proposed by Cornell and Scanlon[6] for new buildings. It is reassuring, however, that their target annual energy demand figures for buildings basically similar to those studied in 1958–59 corresponds so closely to the yardsticks then established from a quite different approach to the problem.

There is room for argument as to the most effective way of expressing energy targets, and whether design targets or annual energy targets are the more effective in exercising the necessary discipline over energy consumption.

So far as centrally heated office blocks are concerned, it would appear that annual energy targets will be more effective once it becomes accepted that walls (and the percentage of glazing permitted) fall in line with the new Regulations for domestic buildings (or some other standard). Acceptance of this control and of the rate of air change, automatically sets the 'design target' by determining the rate of heat loss to be made good by the heating system. The only thing that then matters in the long run is the energy consumed per unit of heat requirement—GJ/kW or MJ/w. The target is then capable of being applied directly irrespective of shape, orientation and size of the building concerned, as long as the heat requirements have been calculated. There is thus some advantage in expressing annual energy targets in this way and this is how the IHVE Guide 1970[7] refers to 'its useful check figure for intermittently heated buildings' as being 5·4 GJ/kW of installed capacity. So now if the 3 different ways of expressing annual energy targets are compared we find:

Knight and Cornell[1]	6·05 MJ/w
Cornell and Scanlon[6]	6·13 MJ/w
IHVE Guide[7]	5·40 MJ/w

The IHVE Guide[7] figure is based on installed capacity including standby. It is reasonable to assume this figure should therefore be increased by 10–20 % to be directly comparable with the other two. This is clearly a very close correlation indeed.

Similarly, there is scope for discussion on whether annual energy targets should be expressed in terms of net energy input to the building or as gross or primary energy demands. Certainly there is a need for defining the terms used. Gross or primary energy demands are useful in comparing the effects of using the various energy sources as is particularly the case when comparing the performance of air conditioned buildings. Designers of air conditioning have the opportunity of deciding how the various energy sources shall be employed and can opt, for example, to change from electrically driven refrigeration compressors to steam or gas fired absorption plants or to diesel engine driven compressors, all with widely different effects on primary energy consumption. However, the choice facing designers of centrally heated buildings usually is limited to the fossil fuels for firing the boiler plant and electricity for pumps, lighting and lifts. In such cases there is much to be said for stating annual energy targets as net energy inputs to the building—useful energy to the building owner. Otherwise, confusion can easily occur as is apparent when note is taken, for example, of the recent paper on Energy Consumption in Tall Office Buildings by Milbank[8] of BRE. He suggests that on average the 15 m wide building with natural ventilation needs 2·15 GJ/m^2 floor area per year. He also considers that by using better controls the figure could be cut to 1·80 GJ/m^2.

On the face of it these figures seem to be very different indeed from those previously discussed, and not accounted for entirely by the fact that Milbank's[8] are in terms of primary energy while the others are net. An analysis of the way in which the figure of 2·15 GJ is arrived at, however, shows that 0·84 is included as electricity for lighting. Also that 2500 hours of artificial illumination is taken against Cornell and Scanlon's[6] 1125 hours, although W/m^2 is the same.

An allowance of 10 % is made for distribution losses in supplying the fossil fuel. Cornell and Scanlon[6] include a 5 °C gain from fortuitous sources. Milbank[8] ignores this and assumes 1800 full load hours of operation against 1600. There are a number of other differences in approaching the problem but when the Millbank[8] figures are adjusted to give net energy into the building and the same assumptions as to hours of operation etc. are made, the figure 1·3 GJ for heating out of 2·15 GJ becomes only 0·77 GJ. Now a direct comparison can be made with the 3 other sources. It can be deduced from the annual energy demand of 0·77 GJ m^2, and what is known of percentage of window areas in the Milbank[8] cases, that these existing buildings would have design fabric loss targets of about 47 W/m^2, a ratio according to Cornell and Scanlon[6] of 4. For this type of building according to the tables of energy demand the annual energy consumption target would be about 320 MJ/m^2 and thus, 6·8 MJ/w. The apparent great discrepancy in the two ways of expressing energy demand can, therefore, be explained, and once again there is strong correlation of the targets separately arrived at. To the party concerned with obtaining quick guides to energy consumption of buildings there is, however, a real danger in making comparisons out of context. It would seem to be better if energy demands for centrally heated buildings were expressed in terms of net energy input. The argument is, however, probably not so readily supportable when considering the overall economics of energy demand of domestic dwellings, factories and air conditioned buildings. Preparation of such targets for all the various types of heated buildings, will be of great value to future designers and allow owners to see the penalties in energy demand terms of departing from the most economical type and shape of building. So might be created the climate of energy planning which has been lacking in the past.

A refinement of these energy target figures will be necessary if we are to take into account, as eventually we must, the overhead and distribution losses of the various forms of energy. This will be particularly important when attempting to make comparisons of air conditioned and other buildings where a number of energy choices are available to the design engineers as mentioned earlier. From all these considerations the only way to make valid comparisons in the national interest is on the basis of the primary energy used to produce the net results. It is the lack of this basis for comparison that has hindered our appreciation of the impact on energy demand presented by the modern air conditioned office. The lack of energy demand targets occurs because of the complexity of making valid comparisons when there are some twelve different types of air conditioning systems all differing in their ability to achieve the seven or so different functional requirements of air conditioning. A variety of choices governing the type of refrigerator machinery and condenser cooling systems: of air and pump handling capacity and whether all-air, air-water; central or localised plants—not to mention wide differences in capital cost—make it virtually impossible to produce the design targets referred to earlier, especially so if they are to be applicable to all building shapes and wall-window ratios currently used.

Perhaps if and when a 'feed-back' of the records of most of the air conditioned buildings now in use is made available, it may be possible to produce some average rates of energy used by the different combinations of plant and so establish a guide for the future. This has been found possible in government housing in the tropics, where air conditioning is justified by the severity of the climate, and where like can be compared with like so far as the user, building and system is concerned. In the meantime, the only way to properly assess the energy demand for a particular air conditioned office building is to make a comparison of the different ways of achieving the same overall result. This can be done by drawing up a balance sheet of total energy demand of all the component parts, (fans, compressors, pumps, boilers, heaters and so on) for the systems considered as capable of fulfilling the functional requirements. This then has to be read in conjunction with a similar assessment (or guarantee) of the raw input energy consumed over an average full year's operation. The results will be similar to the capital and operating cost analysis, usually made to justify further capital expenditure and even carried out in a kind of discounted cash flow manner. While producing more work for designers such a systematic approach to decision making will not merely save money in the short term, but lead to a general improvement in the quality of the design effort and in the equipment used. The increased demand for energy as a result of growth in the rate of supply of air conditioned office accommodation will thus be kept within reasonable limits.

REFERENCES

1. Knight, J. C. and Cornell, A. A. (1959). Degree days and fuel consumptions for office buildings. *J. IHVE.*, **26,** 309.
2. Knight, J. C. and Jones, L. L. (1963). Fuel consumption of off peak floor heating compared with continuous and intermittent central heating systems, *J. IHVE.*, **31,** 81–299.
3. Harrison, E. The intermittent heating of buildings. *J. IHVE.*, **24,** (1956) 145 and *J. IHVE.*, **24,** (1957) 500.
4. Colthorpe, K. J. (1964). Intermittent heating of light structure buildings. *J. IHVE.*, **32,** 165.

5. Jackson, B. Optimum Start Control. *J. IHVE.*, **39,** (1971) A24 and *J. IHVE.*, (1971) A34.
6. Cornell, A. A. and Scanlon, P. W. (1975). Energy targets for office buildings and barrack blocks. *J. IHVE.*, **43,** 118.
7. *IHVE Guide Book A* (1970) p A9–2 to 8.
8. Millbank, N. O. (1974). 'Energy Consumption in Tall Office Buildings,' Conference of International Association of Bridge and Structural Engineers (British Group) and the Institution of Structural Engineers—Tall Buildings and People?

14

Energy Prospects

PETER J. JONAS

Synopsis

The role of Government in the production and efficient use of energy is summarised, and the economic and social consequences of present and future policies, with particular reference to the nationalised fuel industries are examined. The current and projected pattern of fuel consumption in the domestic sector and the means of fulfilling future energy demands from conventional and as yet undeveloped sources of energy is discussed. It is suggested that hitherto apparently unlimited supplies of cheap energy have inhibited the financing of research and development on the efficient use of conventional fuels, and that considerable commercial returns are still possible from the improved design and operation of existing equipment.

Introduction

World energy demand has been growing exponentially at around 5% p.a.[1] Recent events may have reduced this somewhat, but whilst people and nations expect to become more wealthy, economic growth will continue, and with it, energy growth. Estimates may now have declined to around 4% p.a.,[2] but in the UK, the historic rate has been around 1·7% p.a.,[3] with economic growth at around 2·7% p.a.[3]

Although the UK is fortunate in being able to look forward to energy self sufficiency within the next few years, we are not insulated from the international scene. Most developed countries are energy importers, and as fossil fuel reserves are used up, prices will rise and we will have to find alternative sources. Many predictions[4] suggest that the peak of oil and gas production both internationally and in the UK could be passed during the 1990s or before. There is thus a problem, and the Government energy conservation campaign, together with new developments being encouraged by the increased emphasis now being placed on energy research and development will, it is hoped, ensure adequate supplies for the foreseeable future, though perhaps at considerably higher cost.

All the above material is fairly common ground, but the conclusions which can be drawn are often conflicting. For many, the solution will follow from market forces, largely unaided by Government intervention. For others, a strong and restrictive energy policy is necessary to ensure that future generations will be able to rely on and enjoy higher living standards, without disturbing the balance of energy demand and supply. In either case, domestic consumers and commercial or industrial managers responsible for space heating, who control over half our fuel use, will be in the front line with difficult choices and decisions and a welter of contradictory influences.

Many of the same difficulties are also facing the Government, and the Department of Energy is naturally expected to offer advice and solutions. The object of this paper is to describe some of these considerations, and especially those which are particularly relevant to energy use in buildings.

Policy Considerations

A good starting point is Section 1 of the 1945 Fuel and Power Act,[5] which lays on the Secretary of State the duty of 'securing the effective and co-ordinated development of coal, petroleum and other minerals and sources of fuel and power in Great Britain, of maintaining and improving the safety, health and welfare of persons employed in or about mines and quarries therein (now transferred to the Health and Safety Executive) and of promoting economy and efficiency in the supply, distribution, use and consumption of fuel and power, whether produced in Great Britain or not'.

In practice, this philosophy can be broadly translated into the following general objective: 'to minimise the national cost in real resources of supplying the country's energy requirements over time and at generally acceptable levels of supply reliability, subject to wider economic and social considerations, including aspects of national security, the environment, and the often undesirable local effects of rapid change.'

One difficulty lies in finding the right kind of mix of freedom of choice for consumers and energy producers on the one hand, and Government interference, regulation and subsidy on the other. Consumer freedom of choice is of course one of the mainstays of our way of life. In a predominantly market economy we rely to a very large extent on the interplay of demand and supply, influenced by costs and prices, to maximise welfare and ensure an appropriate allocation of resources. In the real world, the forces of free competition and the mobility of factors of production are heavily constrained, and this is particularly true in the energy sector which is subject to natural and institutional monopolistic features, long time lags, and technological and other rigidities. The Government therefore has to establish a framework of guidelines and operating principles within which the fuel industries, and particularly the nationalised industries, have to operate. While recognising the jealously guarded independence of these bodies, the pressures of national and consumer interests have to be built into the background of their activities.

Part of the general background framework consists of the following:

(a) Financial objectives which ensure an adequate return on investment. Inflation, however, tends to erode financial soundness because accounting practice has hitherto made little provision for increasing asset values. Now that the Sandilands Committee is proposing appropriate financial guidelines, the nationalised industries will need to reconsider their position, and prices might need to be increased.

(b) On average, prices and tariffs have in the short run to be set so as to meet the financial objectives. Prices should, however, as far as possible, also reflect costs over time and give the right cost message to marginal consumers.

A problem arises when the long run marginal costs are substantially different from average costs. This situation could arise if, for example, we assume that energy prices will rise substantially in the longer term. Some increase in energy prices is likely, but the effect of future increases on today's price is to some extent modified by the use of a discount rate for translating future expenditure/or savings into today's money values, by means of a present value calculation.

Appropriate discount rates are, however, difficult to assess. Treasury guidance relating to the test discount rate (TDR) for the appraisal of capital projects currently suggests 10%. But it does not necessarily follow that this rate should also be applied to tariff calculations or, for example, depletion policy and there is a strong school of thought which believes that a lower social time preference rate should be used instead. In industry, however, rates higher than 10% are the rule rather than the exception, although these often include inflation, whereas the TDR is net of inflation, tax etc.

(c) A major area of government responsibility is the approval of nationalised fuel industry capital investment programmes. Obviously the first criterion of the soundness of a capital project is the TDR itself. There are, however, many assumptions which have to be taken into account. One example is the growth rate of future energy demand, which is in turn dependent on prices which could vary with the level of new investment. Here the Government has added responsibility for giving guidance about likely future growth rates of the economy as a whole, on which energy growth largely depends. There can be conflict—and it is interesting to speculate on whom ultimate responsibility for securing adequate supplies lies. The nationalised industries clearly have a statutory duty to supply fuel and energy whenever required (except that the British Gas Corporation has recently, by special dispensation, been relieved of this duty in respect of large and industrial supplies). The Government has, in general, to be satisfied that capital projects are in the national as well as the industry's interest.

(d) Another important role of Government relates to safety and the environment. This concerns both the workers within the energy industries and the population as a whole. Planning permission is of course required for any new installation and the energy industries have to justify their planning policy to public inquiries who can probe into the safety and relevant environmental and amenity aspects, including the complicated assessments relating to pollution effects of any energy project.

(e) Since the increase in oil prices, the Government has launched the energy conservation campaign, making use of established advertising media and public relations consultants. Concern for energy conservation inevitably leads to involvement in energy utilisation, which since the aftermath of the last war has been very much left to market forces in the belief that natural self interest on the part of the individual will lead to the optimum use of resources including energy, given that prices adequately reflect costs. Energy conservation and hence utilisation have now become a field of Government activity which can conveniently be summarised under the following headings:[4]

(i) securing so far as is practicable that energy prices reflect the cost of the resources devoted to energy production or acquisition. Given that prior to the crisis energy prices had been restrained in the interest of counter-inflation policy, this has meant very substantial increases in price over the past eighteen months;
(ii) a massive public information effort designed to ease and quicken the response of the market to the new price signals;
(iii) financial inducements such as tax remission on insulation in industrial buildings, and an industrial energy savings loan scheme; and penalties such as higher taxes on motor spirit;
(iv) mandatory measures, such as higher insulation standards in new dwellings and speed and temperature limits;
(v) good housekeeping in both the public and private sectors;
(vi) the setting up of a National Advisory Committee on Energy Conservation under the Chairmanship of Professor Sir William Hawthorne, Master of Churchill College, Cambridge;
(vii) last, but by no means least, research and development which covers a wide field stretching from energy analysis, economic and social studies to technological and scientific endeavour, partly financed directly by Government and partly by the energy and manufacturing industries.

The Government attitude to energy conservation is the result of several rather difficult compromises. In the first place, economic necessity requires that the limit to desirable energy efficiency must be set by financial and not purely energy considerations. It would not be in the national interest to conserve energy if by so doing, we were to use more resources than those preserved by the saving in energy. In the second place, the complexity and diversity of energy use limits the extent to which any kind of effective rationing scheme can be introduced, at least without seriously undermining industrial output or creating a sense of grievance amongst many users whose livelihood depends on the use of varying amounts of energy. Only a vast new army of civil servants could attempt any equitable scheme and the cost would be prohibitive. Government subsidies might be considered, but there are many good causes apart from energy, and economic stringency precludes increases in Government expenditure.

Domestic Energy Consumption

Against this background, we now come to consider trends in energy growth. Statistics are available for industrial, domestic and commercial consumption patterns.[3] For the domestic sector, Fig. 1 shows that since 1968 total heat supplied has risen only slowly, and at times, declined. This surprising result is contrary to what might have been expected from the increase in economic activity, and hence the standard of living during this period. The main factor leading to this

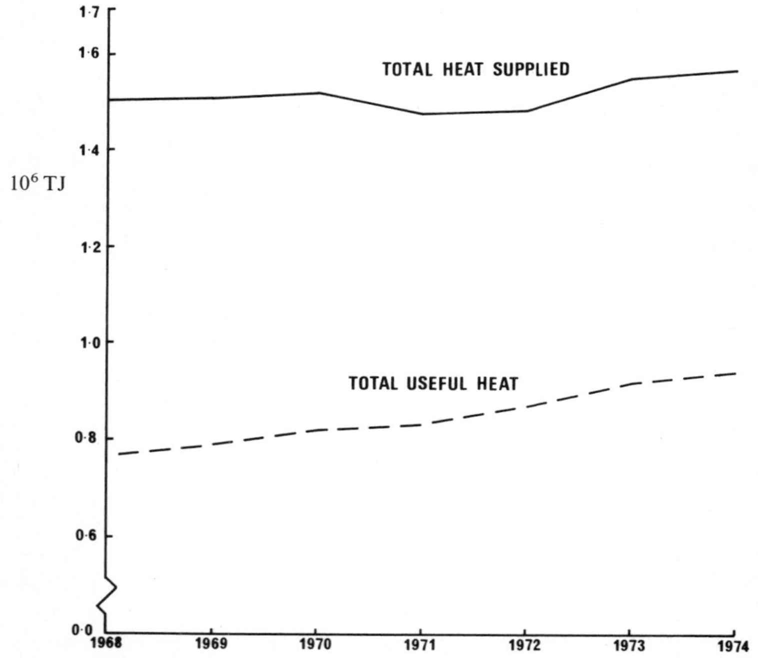

Fig. 1. Domestic energy consumption (temperature corrected for each fuel).

situation is the gradual decline in coal sales, because coal used to be burnt in open grates at much lower thermal efficiencies than gas or in other modern appliances which have replaced open grates. No reliable estimates of thermal efficiencies are, however, available, and any numerical value must therefore be largely based on guesswork.

TABLE 1
NOTIONAL DOMESTIC ENERGY UTILISATION EFFICIENCIES (EXCLUDING PRODUCTION, CONVERSION AND DISTRIBUTION ETC. ENERGY EFFICIENCIES)
TYPICAL HISTORICAL VALUES FOR ILLUSTRATIVE PURPOSES ONLY

	Typical average annual domestic utilisation efficiency
Coal (notional average of open grates with and without back boiler and other appliances)	30%
Other solid fuel (also open grates but a higher percentage of central heating plant)	42%
Gas (mainly modern appliances but making some allowance for summer water heating)	65%
Electricity (including uncontrolled, off-peak storage water heaters)	90%
Petroleum (including free standing paraffin heaters whose efficiency is over 90%)	75%

Important Note
The above efficiency figures have been devised for the sole purpose of demonstrating the difference in growth trends of useful and total heat consumption by final users. For this, they are considered to be sufficiently realistic, but they should not be used in any other context, and they do not in any way constitute Department of Energy estimates.
It will in particular be appreciated that the figures do not allow primary energy inputs to be evaluated.

Typical estimates of average efficiencies are shown in Table 1, although the figures should be taken as purely illustrative of an historical situation. Efficiences of energy use today cover a wide range and are likely to be higher than those shown in Table 1, particularly for coal. The figures do not therefore reflect modern practice and are just one of many possible indications of what might have happened in the past. However, with the aid of such figures we can derive a crude but nevertheless interesting estimate of useful heat. Figure 2 shows individual fuels consumed in the domestic sector and the graph labelled useful heat in Fig. 1 has been derived by adding together the product of efficiency shown in Table 1 and fuel use as in Fig. 2. The important result is that useful heat consumption is rising significantly faster than total heat supplied. The slower rate of growth in heat supplied was due to improvements in overall utilisation efficiency caused largely by the decline in the use of coal. As coal consumption levels out, and the resulting efficiency gain disappears, so total heat supplied will begin to rise in step with the increase in useful heat.

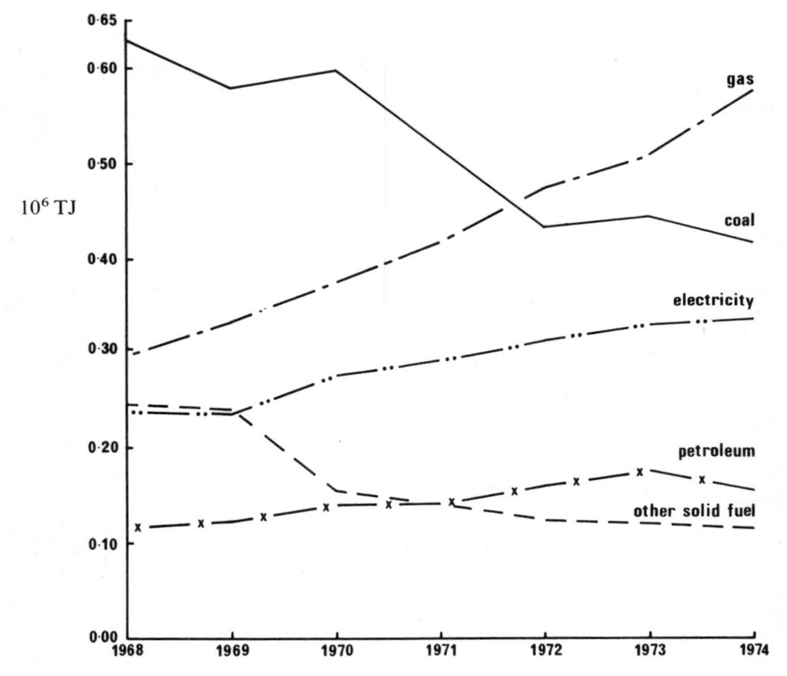

Fig. 2. Domestic energy consumption (temperature corrected for each fuel).

Current indications are that this is already happening, consequently it is now particularly difficult to discern the effects of higher energy prices and of the energy conservation campaign. Their effect is liable to be swamped by the underlying trend of higher heating standards as evidenced by the trend in useful heat consumption. This trend is the result not only of increasing disposable income, but is also, and perhaps substantially, due to improved heating installations in the new and converted dwellings, and due to modernisation and improvement schemes. Even in the present period of economic stringency, these improvements are continuing, so that further increases in domestic heat consumption can be expected. But it seems certain that these increases would have been much higher if energy prices had not been raised, and if the Government had not launched its energy conservation campaign.

Domestic Energy in the Future

Projections of total domestic energy consumption into the future should preferably be based on the trend of useful heat demand, and some estimates of efficiencies which themselves will change in time. When using constant efficiencies, the estimates of useful heat will incorporate some distortion, though the estimates of total heat supplied will not be distorted so long as the time trend in efficiency is reasonably constant.

Energy consumption trends are largely determined by economic activity. For the domestic sector, disposable income, consumer expenditure or more generally the Gross Domestic Product (GDP) could all provide sound basis for trend analysis. Over the past 20 years or so useful energy has grown by about 3% per year compound, but insufficient statistical evidence is currently available to predict with any reasonable degree of certainty how this growth rate would change in response to variations in economic activity and price levels and the energy conservation campaign. Experience to date suggests that total domestic energy consumption is relatively inelastic with respect to energy price changes in general, but interfuel price elasticities are more difficult to discern, partly because there has been a continuous growth in consumer expenditure and simultaneous major changes in fuel availability. For the time being it seems unlikely that 3% historic growth rates will continue to be maintained.

Energy Availability and Prospects

This paper has so far briefly reviewed the Government's approach to energy problems and considered energy use in the domestic sector. Energy in buildings covers a rather wider field, but the essential policy and technical ingredients are similar and clearly distinguishable from those relating to the industrial process and transport use of energy, where very different economic and technical factors are at work such as, for example, environmental and general trade considerations. But all forms of fuel and energy use depend on the availability of energy, and there is competition for energy by different users in the same way as there is competition by producers for the same market.

Although we in the UK are fortunate in being able to look forward to energy self-sufficiency by perhaps 1980,[7] there are strong indications that UK and world reserves of oil and gas will not be able to support historic rates of growth into and beyond the 1990s. If market forces are able to operate without constraint, then there is some hope that energy supply and demand will remain in balance, because the prospect of dwindling reserves will accelerate price increases so that demand will be curtailed, and there will be economic justification for developing alternative energy sources. Market forces alone may not be adequate, and there is an overriding need to avoid the socially and economically undesirable effect of large price increases. All Governments are now adopting a policy of energy conservation, and increasing effort is being devoted to energy research and development, in order to speed up the provision of economical substitutes for oil and gas.

Coal

One obvious answer is coal, particularly in the UK, where there are ample reserves. But even the new plans for the UK coal industry[8] largely provide for maintaining output at current or only slightly higher levels, in spite of the massive investment needed (£600 million) over the next few years. New capacity will mainly offset unavoidable losses in capacity due to exhaustion of reserves amongst our ageing stock of collieries. For the time being it is not considered practical to take expansion further. In the longer term economic considerations will determine if viable new markets, such as for the products of coal conversion processes, can be found.

Nuclear

World wide, the fastest growing energy source is nuclear power. In the United States over 36 000 MW had been commissioned by the summer, which is more than half of the total generating capacity by coal, oil and nuclear power in the UK. In all, 200–300 000 MW of nuclear power is firmly committed and some developed countries are intending virtually to cease ordering any other type of power station.

In the UK our pioneering effort in the 1950s and 1960s is helping to cushion the impact of fossil fuel prices, but there has been some delay and only two new stations[9] are planned for the next few years. Due to a slowdown in electricity growth, however, at least compared to recent forecasts, the need to order other types of power stations has declined.

All current orders for nuclear plant are based on thermal reactors, although it has become evident that, as with oil and gas, reserves of uranium would not be able to sustain world wide historic energy growth rates towards the end of the century, unless a much more efficient use of uranium could become practicable, in for example, fast breeder reactors (FBR). Increasing effort is therefore being devoted to improving the design and safety aspects of FBR's and commercial application could begin in a few years' time, particularly in France where a prototype has been operating successfully for over a year. Large scale ordering cannot however be expected in the near future, and will depend on technical developments, energy growth and public acceptance.

Unconventional Energy Sources

The attraction of most unconventional energy sources is that they are self-renewable or virtually inexhaustible and that the direct energy cost is small or zero. Their capital cost is, however, substantial. For the time being their total cost is higher in most, though not all applications, than that of conventional fossil or nuclear energy.

Some of the potentially most valuable new energy sources are still in the embryonic or infancy stages. In the case of fusion power, development is technically and scientifically difficult, if not impossible with current knowledge. It is in the early laboratory stage, suggesting a lead time of 20–30 years at least before commercial exploitation, for which there is however massive potential scope. Fusion is one of the few new energy sources, apart from breeder reactors which could be developed on a world scale.

Undeveloped sources like waves and thermal gradients in the oceans could be exploited with some development, but have hitherto been considered too expensive and impractical. Further research and development is required to see if costs can be brought down. But even if they can be, the scope will be limited by amenity and geographical factors. For the UK, wave power is, of course, particularly attractive, and initial work is progressing on a feasibility study which should clarify the basic economics.

A great deal of successful and practical development has already gone into solar, wind, geothermal, and tidal power, which are all being used on a limited scale. These are economical where fuel prices are high or where circumstances are favourable and where there is enthusiasm and a pioneering spirit. The scope of these sources is substantial, but not unlimited, and they could make a very considerable contribution to energy supplies. Development is at present concentrated on reducing costs, and interest is on the increase internationally.

Energy for Buildings

The paper so far sets a very brief and general framework for a more detailed consideration of the background and prospects for future energy supplies to buildings. Unfortunately the picture is by no means simple, and two general considerations have to be emphasised.

Firstly, it is necessary to emphasise uncertainty and the inability even for Government and their advisers to make reliable predictions about factors which are outside UK control, like future international oil prices and world wide nuclear developments. It is therefore all the more necessary for all concerned to acquire knowledge, and to have some understanding of the problems facing us, and how we can best respond to them.

The second important message is we are living in times of change. We are not really attempting to cater for all types of crises. Some allowance is made for safeguarding essential services during supply interruptions, but we know little about peoples' willingness to pay for secure supplies, in spite of our experience of people's reaction to electricity cuts. But we do plan for changes according to cost and price messages. Energy consumption patterns are, however, subject to long time lags because of the high capital cost of user equipment. In the past, we have relied on low cost and plentiful energy. In future, energy will be more costly, and we have a long period of adaptation ahead of us. Exhortation like the Government's Conservation campaign can accelerate change, and reduce the harmful effect of high prices by anticipating the action required to cope with them.

Energy Prices

Energy prices have recently been adjusted upwards so that they now more closely reflect costs but with continuing inflation further rises can be expected. This follows a period from the late 1960s to 1974 when, in the interests of counter inflation successive governments have prevented the energy industries from raising prices sufficiently to reflect increasing costs. Such distortions affected domestic more than industrial prices, and this led to some difficult border-line cases, in the commercial or large district heating markets. In spite of these and other legacies from the past, the Government have now reverted to the practice of expecting the energy industries to be financially viable, and future price levels should in general, reflect long run marginal costs. In addition, some upward adjustment to allow for inflation seems likely in due course. Moreover, to the extent that UK oil and gas will eventually have to be replaced by imported fuel or other energy sources, it could be argued that all basic fossil heat costs should be equal, and any variations in prices would only be due to differences in transport, distribution, or transmission costs. The Government's depletion policy for North Sea gas and oil will almost certainly be aimed at maximising the benefit the UK can derive from its own resources, and UK oil prices are likely to continue to be based on international oil prices. Although it may not be desirable in the long run to allow gas prices to remain much lower than this, counter inflation and social factors could outweigh these considerations, and in the long term gas prices cannot be predicted with any certainty.

Uncertainties about future availability of gas, both from the UK and from adjacent sectors of the UK continental shelf, and about world oil prices, make it impossible to give clear guidance about the best fuel for use in buildings. Official advice is to consult fuel and appliance suppliers, heating consultants and any other source of information that will enable the user to obtain best value for money in terms of comfort, fuel efficiency and, of course, cost. As referred to earlier, the Government role is to set the general framework which should ensure that individual decisions freely taken, and therefore furthering the interest of the individual, are also in the national interest. But it is important that the decisions are taken on the basis of a sound knowledge and understanding of the situation.

Long Term Prospects

In the long term, nuclear power could become the most economical alternative to fossil fuel. It could also be made to provide direct industrial heat or even heat for the conversion of coal or water to gas or hydrogen. At this point of time, it is not possible to say when such processes might become economical. For the UK, North Sea reserves may allow longer timescales than elsewhere, and for the time being, there does not seem to be a need to anticipate any major changes to our fuel supplies and hence fuel for buildings. The prospect of a 'nuclear future' has been the subject of much speculation, and is naturally under intensive study by the nuclear and electricity industries throughout the world. It certainly emphasises the need to give careful consideration to the use of electricity in buildings, but it would be prudent to await developments in countries which are not so well endowed with fossil fuel as the UK, and which are currently planning a larger nuclear power programme than the UK. On present prices, 'on peak' nuclear electricity cannot compete with fossil fuels in buildings, but by the end of the century, when, as explained earlier, dwindling fossil reserves may have forced a reduction in supplies, prices may have gone up sufficiently to allow nuclear electricity to make inroads into the bulk heating market.

Coal also requires special mention. It is the only indigenous UK fuel with enough reserves to last well into the next century, and perhaps beyond. But here again, costs and hence price will set the scene. Coal is currently the mainstay of electricity generation and ironmaking. If nuclear energy displaces coal in these markets, it will be because it is cheaper, in spite of increasingly stringent safety and environmental considerations. The coal so displaced could be converted to gas or oil, but economic considerations will again set the balance between the different contenders, which could, towards the end of this century, replace conventional gas and oil.

Last, but by no means least, we come to energy conservation in buildings. For all but the smallest domestic energy user, fossil fuel is nearly twice as efficient overall as electric heating. But high efficiencies can still be achieved by combined heat and power plant, sometimes referred to as total energy schemes. These are of course, highly desirable from an efficiency, and hence conservation point of view, but can only be justified if they can be seen to be economically and financially viable. Current energy prices, and possible future increases, would seem to call for a reversal of past trends when the emphasis was often on cheap initial cost. A well designed and therefore efficient domestic heating installation should become a major feature of all newly built or converted homes. One way to encourage this would be to include heating bills in the advertising and sales material usually provided whenever a house is for sale. Such a practice is already established in certain parts of the United States. I am sure other papers have dealt with the technical and practical aspects of the energy savings potential in buildings, which consume around 50 % of total final energy used. I suspect that the long years of cheap energy have prevented adequate research and development funds being devoted to energy efficiency. Heat exchangers, total energy systems, and the proper use of daylight may not have the glamour and sparkle of solar heat or wind power, but I am convinced that there is still plenty of scope for economically desirable, and commercially viable efficiency improvements in the design and use of conventional equipment, both in existing and in new buildings, large and small. An advance on a broad front is needed to ensure that the UK does not fritter away the valuable resources of the North Sea.

REFERENCES

1. 'World Energy Supplies'. Series J United Nations 1972.
2. 'Chase Manhattan Bank Report' March 1975.
3. 'Digest of United Kingdom Energy Statistics', 1974.
4. Select Committee on Science and Technology (Energy Resources Sub-Committee) Session, 1974–75 (156 ii).
5. Ministry of Fuel and Power Act 1945 Ch 19.
6. 'Use of Discounted Cash Flow and the Test Discount Rate in the Public Sector'. Management Accounting Unit HM Treasury, June 1973.
7. *Development of the oil and gas resources of the United Kingdom.* (1975). HMSO on behalf of Department of Energy.
8. 'Coal Industry Examination Final Report' Department of Energy, 1974.
9. *Nuclear Reactor Systems for Electricity Generation* (1974). HMSO Cmnd 5695.

Discussion

A. W. Catterall (*Lewisham Health District*): The question was asked earlier what incentive can be given to the 50 % of the population who are tenants and not home-owners. They get no incentive at all under the price mechanism.

W. P. Jonas: This is a difficult question for me to answer because it is the Department of the Environment which has ultimate responsibility for Government policy relating to local authorities, including local authority housing which comprises most of the rented accommodation. In general the framework I have described should cover local authorities, because they also make use of conventional criteria so as to minimise overall costs. At present however there is an economic situation which does not facilitate extra public expenditure even if subsidies were justified economically. The Treasury may for the time being consider that in this time of economic recession the first priority is to pull out and get industry moving. Increasing public expenditure would detract from our ability to pull out of the economic trough. But the problems facing local authorities when implementing energy conservation programmes are certainly being taken seriously.

It is most important to recognise the very real difficulties encountered when attempting to ensure optimum allocation of resources between Government Departments, and between Government and the economy as a whole. Discussions are proceeding with the Department of the Environment and with the local authorities and the results will be made public in due course. (See also Department of Energy Circular No. 1/75 issued 25.11.1975).

F. A. Pullinger (*Haden Carrier Ltd.*)*:* I wonder if the Department of Energy underestimates the need to encourage us. We do believe that they know so much more than we do about the statistics. They have the funds and the opportunities to collect and develop such information. I very much welcomed Mr. Jonas' paper and his frankness in exposing the uncertainties made his contribution all the more valuable. If we can be shown the underlying principles behind your thinking, then we are able to base our own advice on them. Collectively we really are very powerful agents for doing what you want. You only have to look at the cost of energy in buildings, which is constantly paraded, to see who are the agents for saving—we are. Now we do need to know what you are thinking—for instance, on the question of a subsidy you say your judgement is that it is not of value in tax terms, to subsidise heavily. We could not make that judgement. Now we know more or less where we are and the direction in which our advice should begin. You remind us of the difficulty of predicting fuel prices. This makes legislation so difficult and improbable, so we can deduce that you are trying to let cost savings be the incentive. I really wanted to say 'thank you' for drawing the curtain aside, but also to encourage you to come back and keep up a dialogue. We must talk again.

P. Le Cheminant (*Department of Energy*)*:* I wonder if I can answer Mr. Pullinger's question. The processes of Government over the years have, of course by tradition been shrouded in mystery, but that tradition is dying fast. The Department of Energy is doing its best to 'stab it in the back' if it is not 'cutting its throat' from the front. Certainly I think it is true to say, that in the 19–20 months that we have existed, we have published far more information of a kind that is a basis for discussion of policy than ever before. For example, we have started a series of Energy Papers which have a bearing on future aspects of the energy scene, but which do not in themselves commit the Government to a particular attitude or a particular policy. On Energy Conservation we have issued a pamphlet containing Eric Varley's speech to the Science Writers Association last year, which sought deliberately to set out what the philosophy of our approach to energy conservation was.[1] You also mentioned The Select Committee. In their Report[2] you will find many pages from the Department of Energy of written evidence on our forecasting techniques, our forecasting models and our approach to the problems of uncertainty which totally dominate the energy scene. The basic fact which underlies energy policy is that we do not know what the future holds, so one has to try to devise and use all the available modern managerial techniques for making decisions in such uncertain situations, for example, complex computer simulations. I really would like to express my thanks to Mr. Pullinger for his remarks. It is our wish to publish as much as we can, and I would totally agree that we have not published enough yet. There is more we should do. However, we have started out with what we would hope you would think are the right intentions and we will endeavour to keep it up over the months ahead.

P. J. Jonas: Obviously I agree with much of what has been said, and I think that Mr. Pullinger is now aware of the present position. I want to make it quite clear that I have been speaking about the past and the present. Whether there will be grants or subsidies in the future is something that only the Government and ultimately Parliament can decide. The task of professional advisers is to quantify the costs and benefits, we obviously cannot commit future Governments' action. Any such commitment would be interpreted as being contrary to the principles of a free and democratic society.

J. W. Weller (Adams Green & Partners): Energy Conservation and Energy Management in Buildings are of interest to all occupiers of buildings. Unfortunately the Government's Campaign to 'Save It' (which is presented as a command and not a request) does not recommend the quantities which they would like saved.

Both occupiers and building designers need guidance and I hope that the relevant Government Departments respond to this conference's requests for guidance.

Irrespective of whether this guidance is forthcoming, conscientious Building Design Teams are already becoming more involved in Energy Management and the associated complex analysis of energy movement in buildings. Due to fluctuating internal and external conditions the analysis of energy movement is beyond hand calculation and the design team have to use computers to run programmes.

1. Association of British Science Writers 26 June 1974 Information Division Department of Energy.
2. Science and Technology Energy Resources Sub Committee, (1975). *Minutes of Evidence* 115 viii HMSO.

P. Le Cheminant: Mr. Weller talked about 'Save It'* being a command—it is not a command. Perhaps 'Switch off something now'* was nearer to a command because we had an immediate and actual crisis to cope with. If you want guidance as to the techniques to decide whether to save energy or not ask your accountant for the relevant Test Discount Rate and apply it.

Dr. P. V. L. Barrett (ICI Insulation Service Ltd.): The Department of Energy is going for exhortation. As far as the private homeowner is concerned there is one area of finance which does not seem to have been advertised. We have all heard in this conference of the tremendous 'pay-back' periods. The Government could encourage the Building Societies to use their funds to finance the more expensive forms of insulation such as cavity fill. That would be a good investment by the Building Society and a good investment for the nation.

P. J. Jonas: I think that is a sound suggestion, but I am not sure whether we have approached the Building Societies on this. Certainly we would encourage investment in cavity fill, and we have said this in all the publicity. Double glazing may, however, have a longer 'pay back' period as some of the papers have already pointed out.

May I now go back to the problem of targets, raised earlier. I do not think that it would be easy to say to people 'this is what you have to do' without infringing on their freedom too much. We cannot say that houses must be insulated all at the same time. Maybe the time will come when we have to extend legislation to existing houses, but we do not believe this is the right thing to do at the moment. Nor can we see a realistic way to check up whether the millions of people who use energy would ever achieve their target. It would be futile to set targets without any means of implementing them or enforcing them. This is a good example of the dilemma of conflicting interests we face all the time and the Government has to strike the right balance between exhortation and force.

W. J. Ablett (Ministry of Defence): First of all I should like to say that the 'Save It' label my son stuck on my bedroom door was most effective. There is one point I have been repeatedly stating for at least 35 years and, once again, I see that electricity heating has a 90% utilisation efficiency. I think we should condemn this form of presentation. We know our power stations are about 28% efficient and if it is a coal fired station we know that we have also an energy input to get the coal to the power station. We are therefore saying that the efficiency of that heating appliance is 90% of 28% which is a totally different picture. During these many years I have seen no reason why the domestic consumer should not be put on a maximum demand tariff in the same way as an industrial consumer. The effects of this would in fact be to make the electric fire and electric cooker almost obsolete. We have been told

* Editor's Note—Both the phrases quoted have been heavily used in government advertising in the UK 'Switch off something now' during the energy crisis of January–March 1974 and 'Save It' during 1975.

that there are roughly 18 million homes in the UK, of the 18 million I would hazard a guess that 5 million have these electrical appliances. If one works out the total load of these, it is equivalent to one power station, and having recently read that one power station is now costing £1200 million, has been 8 years in construction and has still to produce a unit of electricity, it would seem to me that there is some money somewhere to carry out energy conservation measures. One further comment—in the domestic sector electricity tariffs have recently had an added fuel clause. We are going in the right direction.

P. J. Jonas: I am glad that you have raised the question of efficiency. The footnote in Table 1 does in fact refer to the difference between utilisation and generation efficiency. The table was produced entirely to assist in forecasting energy demand, when one has to estimate useful energy demand. If we are looking for the cooker load for example in the domestic sector, then we have to take the heat in the electricity that is actually used. We cannot look at the power station input in order to estimate what the domestic useful heat consumption is going to be in 10 years time, but I accept the point completely that these figures should not be used to assess the primary energy used for the domestic sectors, only the useful heat that we are concerned with. Having established the domestic use of different forms of energy, the resulting primary energy requirement can be assessed according to generation efficiencies etc.

P. Le Cheminant: I think it is absolutely essential that prices reflect real cost then if somebody chooses to cook on an electric cooker or use an electric fire, that is their business.

A. P. Oppenheimer (*Leicestershire County Council*): We have had some excellent speakers at this conference, but one was missing—one of our great grandchildren. He would come along—that is if he could find enough fuel for his time machine—and he would say 'I am glad you have met here today and I think you were very sensible to look at the short-term aspects first' but then he would add 'we have no man-made fibres, no plastics and no lubricating oils because you burnt most of the basic hydrocarbons'. Should we give more emphasis to research for the generation of electricity, from solar energy, perhaps in the Sahara, from the sea or from other means not involving hydrocarbons. Secondly, should we think about pricing fuels in relation to future need rather than their cost to-day?

P. Le Cheminant: Your great grandson would certainly want to say some of those things. He would not say it quite that starkly. Even in his day there would still be an awful lot of coal around which could provide a chemical base for plastics. The price may be different and we may have had a certain amount of trouble persuading the Oxford Dons that it is a good idea to have massive open-cast coal workings in Oxfordshire—just as we may have had a certain amount of trouble ripping up South-Eastern England for oil-shale. They are both there and the day may come when we need to exploit them and I wish the then Department of the Environment well of the planning inquiries! As to the other sources of energy, I wholly agree it will be delightful if only we could suddenly switch ourselves in to the renewable sources and solve a great many problems. I do assure you that they are not forgotten. There is a lot of work going on partly in this country, but more importantly world-wide and we are nowadays linked into it through the International Energy Agency on the questions of solar energy, wind power, wave power, tidal power, geothermal etc. You know the future is not entirely bleak and it is not entirely nuclear though there is bound to be, we think, a very substantial nuclear slice in the future which will, of course, give us our electricity. If one uses that electricity, or some of it, to make hydrogen and uses the lime-stone, which is fairly abundant—perhaps we might even find a route there to making plastics.

W. R. H. Orchard (Orchard Partners): I was most interested to hear that it appears to be Government policy that fuels are going to reach their market price levels. I would ask how long this is going to take because it would appear that Government is still restraining the fuel industry's price increases. Local Authority housing provides an anomalous situation where for an efficient central boiler plant running at between 80–86%, with possibly 15% total losses for distribution, the gas supplied costs more than the gas supplied at the marginal cost rate to a domestic boiler on the 'Gold Star' tariff, running at a very debatable average annual efficiency of 53% and below.

In the long term it would appear that gas is basically the more useful fuel, either as a fuel or as a chemical feedstock, than either oil or coal. Both for industrial applications and for heating, gas can be used in circumstances where oil or coal cannot, on account of either the sulphur content or storage, or ash handling problems. Gas therefore is a more flexible fuel and yet currently we have the situation where gas is cheaper than oil.

In considering Local Authority dwellings and the provision of heating and hot water for these dwellings I would suggest that they are not really properly part of the private sector, where market forces should operate if allowed to by the Government, but the public sector. In this country we enjoy part private and part public sector activities, and in the public sector it would appear that a different form of political economics is used to evaluate projects, which quite frankly I do not always understand. I therefore suggest that these different political economics be used to good advantage for once, and that for public sector schemes the Department of Energy give those of us who are evaluating the long term capital and running costs for these types of schemes, comparative fuel costs to use in our evaluations, which take into account the long term interests of this country.

P. Le Cheminant: First of all I think we have to recognise the distinction between getting rid of subsidies and realistic pricing. The Government is committed to end subsidies during this financial year which is a reason for the recent price increases. Part of the problem is that we are constrained by law. For example, the Price Code is a cost related code and, for example, the price of gas is not what the gas industry could screw out of the market in the last resort if it were given its head, but what it actually costs to produce. It is of course very relevant to try and ask oneself what is a realistic price, is it related to the costs of the industry *per se*? Is it related to the costs of the industry as they would be if the Sandiland Committee's recommendations[1] on accounting procedures in an inflationary situation were taken into account? Is it thermal parity, or what? One can set a whole set of levels and say that this is a real price and so is this, even though they are different. Nevertheless I think it is true to say that a great deal of progress has been made in the last 12–18 months in removing the distortions between the different energy prices, which arose in part from quite deliberate counter inflation policies which were perfectly sensible in their context and in part from the very dramatic increase in the price of oil. Both these things threw distortions into the system which have mostly now gone. We have got nearer to thermal parity in prices. Though we are not at thermal parity which is anyway a difficult thing to define.

P. J. Jonas: One particularly difficult problem affects gas pricing. Here we are not necessarily dealing with a commodity like oil where you can turn the tap on and off. The Gas Corporation has to enter into long-term contracts with the suppliers of gas from the North Sea, including of course non-UK suppliers like the Frigg Gas Field and supplies of gas associated with oil fields, which have a high load factor. These will in due course make a very substantial difference to our gas supplies and hence consumption. Now there is a problem here which it is almost impossible to solve without infringing some of the criteria which we would like to adhere to. One of the criteria for pricing that the Gas Corporation has to adopt, given that they have to meet a given supply target (this is a different type of target to what has been discussed earlier) is that they have to meet a given supply constraint, which they cannot go below without incurring a substantial penalty, and that is the reason why they in some cases have to fix their prices according to what the economists call a market clearing price which could of course conflict with other pricing rules which we might wish to have for conservation reasons. But these factors are largely temporary because once the market has grown and the general level of fuel consumption increases in line with (as I said earlier) a general increase in Gross Domestic Product and wealth, these prices can be adjusted according to any more important conservation criteria. In the past there have been some conflicting messages, particularly due to the Government's counter inflation policy which has led to some underpricing, but this policy is no longer being pursued.

J. Peach (*Institution of Heating and Ventilating Engineers*)*:* I should like to mention a few small points. We have heard quite a lot during this conference about building insulation, but very little about personal insulation. Reductions in room temperature can make quite sizeable savings in energy consumed and by adjusting one's clothing one can experience the same feeling of comfort over a range in temperature of 3 °C or more. A reduction in temperature of 3 °C in a heated building could lead to a 20 % saving in energy use. Personally, I am not particularly enamoured of long underpants and

1. *Report of Sandiland's Committee* (1975). Cmnd. 6225 HMSO.

ex-RAF flying boots and I am wondering whether the Department, as part of its Energy Conservation programme, has been in contact with the clothing designers and manufacturers with a view to them marketing a range of conservation clothing for those who might like to use it.

Mr. Jonas mentions that the North Sea oil will give this country breathing space compared with some of our neighbours. I hope we are not just going to sit back during that period. I would hope that our manufacturers will be considering the design of equipment in connection with alternative energy sources. Although during this forecast period of energy self sufficiency we may not need to introduce alternative energy sources for our own needs, the equipment might at least form a more profitable export than some other products which we endeavour to sell, and in the end we shall need both the know how and the hardware for our own use.

P. J. Jonas: Regarding the use of North Sea reserves I agree that we are all facing a challenge. Our objective is to derive the maximum advantage from North Sea reserves. Obviously we must not allow these resources to be frittered away inefficiently and ineffectively or we face the prospect that we will be no better off when they are exhausted.

Index